Gender-Structured Population Modeling

FRONTIERS IN APPLIED MATHEMATICS

The SIAM series on Frontiers in Applied Mathematics publishes monographs dealing with creative work in a substantive field involving applied mathematics or scientific computation. All works focus on emerging or rapidly developing research areas that report on new techniques to solve mainstream problems in science or engineering.

The goal of the series is to promote, through short, inexpensive, expertly written monographs, cutting edge research poised to have a substantial impact on the solutions of problems that advance science and technology. The volumes encompass a broad spectrum of topics important to the applied mathematical areas of education, government, and industry.

BOOKS PUBLISHED IN FRONTIERS IN APPLIED MATHEMATICS

Iannelli, M.; Martcheva, M.; and Milner, F. A., *Gender-Structured Population Modeling: Mathematical Methods, Numerics, and Simulations*

Pironneau, O. and Achdou, Y., *Computational Methods in Option Pricing*

Day, William H. E. and McMorris, F. R., *Axiomatic Consensus Theory in Group Choice and Biomathematics*

Banks, H.T. and Castillo-Chavez, Carlos, editors, *Bioterrorism: Mathematical Modeling Applications in Homeland Security*

Smith, Ralph C. and Demetriou, Michael, editors, *Research Directions in Distributed Parameter Systems*

Höllig, Klaus, *Finite Element Methods with B-Splines*

Stanley, Lisa G. and Stewart, Dawn L., *Design Sensitivity Analysis: Computational Issues of Sensitivity Equation Methods*

Vogel, Curtis R., *Computational Methods for Inverse Problems*

Lewis, F. L.; Campos, J.; and Selmic, R., *Neuro-Fuzzy Control of Industrial Systems with Actuator Nonlinearities*

Bao, Gang; Cowsar, Lawrence; and Masters, Wen, editors, *Mathematical Modeling in Optical Science*

Banks, H. T.; Buksas, M. W.; and Lin, T., *Electromagnetic Material Interrogation Using Conductive Interfaces and Acoustic Wavefronts*

Oostveen, Job, *Strongly Stabilizable Distributed Parameter Systems*

Griewank, Andreas, *Evaluating Derivatives: Principles and Techniques of Algorithmic Differentiation*

Kelley, C. T., *Iterative Methods for Optimization*

Greenbaum, Anne, *Iterative Methods for Solving Linear Systems*

Kelley, C. T., *Iterative Methods for Linear and Nonlinear Equations*

Bank, Randolph E., *PLTMG: A Software Package for Solving Elliptic Partial Differential Equations. Users' Guide 7.0*

Moré, Jorge J. and Wright, Stephen J., *Optimization Software Guide*

Rüde, Ulrich, *Mathematical and Computational Techniques for Multilevel Adaptive Methods*

Cook, L. Pamela, *Transonic Aerodynamics: Problems in Asymptotic Theory*

Banks, H. T. , *Control and Estimation in Distributed Parameter Systems*

Van Loan, Charles, *Computational Frameworks for the Fast Fourier Transform*

Van Huffel, Sabine and Vandewalle, Joos, *The Total Least Squares Problem: Computational Aspects and Analysis*

Castillo, José E., *Mathematical Aspects of Numerical Grid Generation*

Bank, R. E., *PLTMG: A Software Package for Solving Elliptic Partial Differential Equations. Users' Guide 6.0*

McCormick, Stephen F., *Multilevel Adaptive Methods for Partial Differential Equations*

Grossman, Robert, *Symbolic Computation: Applications to Scientific Computing*

Coleman, Thomas F. and Van Loan, Charles, *Handbook for Matrix Computations*

McCormick, Stephen F., *Multigrid Methods*

Buckmaster, John D., *The Mathematics of Combustion*

Ewing, Richard E., *The Mathematics of Reservoir Simulation*

Gender-Structured Population Modeling

Mathematical Methods, Numerics, and Simulations

M. Iannelli
University of Trento
Povo, Italy

M. Martcheva
University of Florida
Gainesville, Florida
USA

F. A. Milner
Purdue University
West Lafayette, Indiana
USA

siam
Society for Industrial and Applied Mathematics
Philadelphia

10 9 8 7 6 5 4 3 2 1

Library of Congress Cataloging-in-Publication Data

Iannelli, Mimmo.
Gender-structured population modeling : mathematical methods, numerics, and simulations / M. Iannelli, M. Martcheva, F.A. Milner.
p. cm. -- (Frontiers in applied mathematics)
Includes bibliographical references and index.
ISBN 0-89871-577-6
1. Sex distribution (Demography)--Mathematical models. 2. Population--Mathematical models. I. Martcheva, M. (Maia) II. Milner, F. A. (Fabio Augusto), 1954- III. Title. IV. Series.

HB1741.I16 2005
305.3'01'5118--dc22 2004065084

Contents

List of Figures

List of Tables

Preface

For thousands of years humans have been occupied, and even preoccupied, with counting. As the species evolved, an interest grew in knowing and forecasting the size of various populations, such as animal herds that humans hunted for their food and clothing or crops their diet depended on.

The first mathematical model for populations is attributed to Leonardo Pisano, better known as Fibonacci, who described in a publication dated 1208 the cumulative size of a population of rabbits after n successive generations through his celebrated sequence

$$s_0 = 1, \quad s_1 = 1, \quad s_{n+1} = s_n + s_{n-1} \ (n \geq 1).$$

Among many interesting properties the sequence has, it is nice to note that the ratio of consecutive terms approaches the golden ratio $\frac{1+\sqrt{5}}{2} \approx 1.618$. This leads to an exponential growth of the population at an asymptotic rate of 1.618 per generation.

Most of the models proposed since then are concerned with a population that may be unstructured, or may be structured according to one or more important features such as age, sex, race, size, and economic status.

For human populations it is quite useful to have models that are structured at least by sex and age, because many health care, education, and social security issues, for example, depend on the sex and age structure of the population. Yet few studies of the mathematical properties of such models have been done and, in particular, a major gap still exists in the study and modeling of marriage functions.

Among all the processes intervening in the dynamics of human populations, perhaps migration is the least understood and most difficult to model. In this book we shall ignore it altogether and assume that the population studied is closed in the sense that individuals arrive into it only by birth and leave it only by death. Processes related to births and deaths are among those we can understand and model with the greatest ease and, when reliable data are available, the models lead to fairly accurate predictions of the evolution of the population—even when modeled linearly. In terms of a sex-structured model, this means that the equations used for the dynamics of the age structure of females and males can be deterministic, and the error stemming from real-life randomness can be neglected.

Equations that model couples' dynamics become necessary in order to have some closed form for the birth functions of females and males. Ignoring couples and applying linear extrapolation of known birth data leads to significant errors by the tenth year—when a new census is taken and new data are thus acquired to compare prediction and reality.

This creates a need for modeling couples separately, allowing the nearly linear birth processes to be described in terms of the age distribution of couples and their fertility, thus resulting in more accurate long-term projections.

If we choose to model couples using differential equations, these must include terms corresponding to divorces, separations, and marriages. The dynamics of divorces and separations is fairly well understood and is actually modeled very similarly to the death process, *linearly*, without the introduction of large errors. Marriages are much more complex. Assuming that they are constant leads to very poor estimation of births from married couples and, therefore, should be avoided. In fact it is always expected that the mating process is nonlinear in terms of available "singles."

In this book we concentrate our attention on deterministic models for monogamous populations, in which the past history of marriages and divorces plays no role in future behavior with respect to these processes.

The problem of existence of a marriage function and its form began being widely discussed sometime in the early 1970s. It was interesting both from the point of view of the two-sex model, because it was expected to help make sense of male and female marriage rates, and for the possibility of forecasting marriages, for different purposes—mainly business related. As demographers define it, a *marriage function* is a function for predicting the number of marriages that will occur during a unit of time between males in particular age categories and females in particular age categories, from knowledge of the numbers of available singles in the various categories.

Marriage is a complex socioeconomic process influenced by many factors, just some of which are dominating perceptions and rituals, religion, health, economy, educational status, racial and ethnic interactions, and age and sex composition. As can be seen from the definition of the marriage function, it is assumed that age and sex composition are the only essential properties that should be explicitly taken into account in modeling marriages.

A general perception in demography is that the marriage system represents a market and is ruled by laws of competition. At a personal level marriage is an act expected to bring more comfort in life—health and/or premarital childbearing are some of the factors that can decrease a person's chance of getting married.

Comparative studies relative to actual population data involve few of the known candidates for marriage functions and we frequently find a weighted harmonic mean as the function of choice in mathematical models, even though demographers know that it is not a good choice.

We try to present in this book a brief historical perspective of deterministic modeling of human populations and then focus on pair formation (marriage) and two-sex models. We describe several models, derive theoretical results that show these are well posed, and try to elucidate which marriage function might make a better choice for a particular population—in our example, that of the United States. We describe numerical methods to approximate the solution of the differential models, which is equivalent to creating discrete simulators.

We present comparisons of simulation results with actual demographic data in the hope that they will help the believer better understand some of the difficulties concerning the availability of data and demographic modeling, and that they might convince the skeptic that mathematical demography does provide reasonable qualitative and quantitative estimates.

Acknowledgments

The authors would like to thank Carlos Castillo-Chávez for first suggesting to Maia that this book should be written and for his enthusiasm about its future.

Maia's doctoral dissertation under Fabio's supervision was the seed for this book: as she worked on it, Mimmo spent several months at Purdue University and the three of us had innumerable discussions on the topics of marriage and sex that are central to this book. Several results from such discussions became publications in scientific journals, others just opened our eyes to problems we could not solve at the time, yet others gave us new insight into the marriage problem—as well as into the problems of marriage.

This book developed and came into being as the result of many years of friendship among the authors and through constant encouragement and support of each other. We are grateful for the time we spent together—especially during the final phase of writing when we all were together for over two months in Trento, from April to June 2003—but also for many pleasant times shared through the years by all our families. These were times of very hard work, but also camaraderie, fun, and friendship that we will cherish all our lives.

We want to express our thanks to the Departments of Mathematics in Trento and at Purdue University for the hospitality and support provided for the three of us. We also thank the National Science Foundation and the Istituto Nazionale di Alta Matematica for their financial support during part of the time in the form of specific grants and contributions.

All of us should thank many people and institutions that directly or indirectly contributed to this book. In particular, we thank Robert Schoen for his numerous helpful comments during the final stage of preparation of Maia's thesis, some of which were incorporated in this book, as well as Todd Arbogast for sharing Fabio's initiation into two-sex population models and for writing the first computer program we used for such models. Maia thanks the Chair of the Mathematics Department at Polytechnic University, Erwin Lutwak, for his constant endorsement of her research endeavors—including this book.

We are grateful to our friends and colleagues—Carlos Castillo-Chávez, Odo Diekmann, Karl Hadeler, Andrea Pugliese, and Horst Thieme—for having shared not only scientific information, comments, and opinions but also general views on applied mathematics that have frequently helped us to endure useless arguments with the many Don Ferrantes[1] of our times.

Finally, and most importantly, we are grateful to our families for their patience in bearing our absences in body and in mind, and for the encouragement we received to complete our work. In this respect each of us is indebted to our own family, but also to those of the coauthors. Thus we want to mention them in a unique alphabetic list, to which Fabio has especially contributed to make so long. We thank Daniel, Diana, Eric, Giovanni, Jacopo, Mariaconcetta, Marina, Marta, Misha, Monica, Sasha, Theodore, and Tzvetan: this book is dedicated to them.

M. Iannelli
M. Martcheva
F. A. Milner

[1]Don Ferrante is a character of the Italian novel *I Promessi Sposi* by Alessandro Manzoni. In the novel Don Ferrante dies from the plague that spread in Milan in the year 1630. He dies arguing against the existence of the plague, his opinion being that plague does not exist because it is neither *substance* nor *accident*.

Chapter 1

Historical Perspective of Mathematical Demography

Nove mesi a la puzza: poi in fasciola
Tra sbaciucchi, lattime e lagrimoni
Poi p'er laccio, in ner crino, e in vesticciola,
Cor torcolo e l'imbraghe pe carzoni.

Poi comincia er tormento della scola,
L'abbeccè, le frustate, li geloni,
La rosalia, la cacca a la sediola,
e un po' de scarlattina e vormijoni.

Poi viè l'arte, er digiuno, la fatica,
La piggione, le carcere, er governo,
Lo spedale, li debbiti, la fica,

Er zol d'istate, la neve d'inverno ...
E per urtimo, Iddio ce benedica,
Viè la morte, e finisce co l'inferno.

—G. G. Belli *"La vita dell'omo"*[1], sonnet 745,
January 18, 1833

The "vita dell'omo" (the life of man) has always been a central concern for humankind. Here Giuseppe Gioachino Belli, an Italian poet living in Rome in the nineteenth century, expresses his pessimistic view of existence with an intensive listing of the many important steps in the life of a single man.

[1] Nine months as a fetus: then wrapped in bandages/among kisses and big tears/then on a leash, in the walker, and in little vests,/with suspenders and toddler's pants./Then school torments come,/A-B-C, whipping, frostbite,/rubella and diarrhea,/some scarlet fever and worms./Then work comes, hunger, fatigue,/rent, jail, government,/hospital, debts, sex,/sun in summer, snow in winter .../And, at last, God bless us,/Death comes and we end in Hell.

Actually, demography has the same concern since its goal is the description of human populations—though it may be objected that demography considers the population as a whole, and its main object of interest is the number of individuals rather than the specific worries of any single individual. However, as we shall see through this book, a deep understanding of the growth of a human population cannot really forget the behavior of single individuals and, consequently, cannot disregard the many minor or major concerns that, at the individual level, motivate human behavior and a significant structure of the population.

In this chapter we shall review early attempts to describe the growth of human populations and introduce age structure as a major characteristic of differentiation among individuals. This is a preliminary step to lead us into the introduction of sex structure, which is the main concern of this book.

1.1 From Fibonacci to Lotka

One of the main characteristics of any population is the number of individuals in the population, and since ancient times people have been interested in enumerating populations, both human and animal. Documentation on counting human populations exists as early as 3800 B.C. in Babylonia, but traces of concern with mortality can be found as early as the Stone Age (as cited in [57]). Inventories of people have been taken throughout history. The early ones are characterized by the fact that they were produced to identify taxpayers, or men of the appropriate age for military service. Every five years the Romans enumerated citizens and property to determine taxes. However, most of these records were inaccurate. The first complete record of people and land was the famous Domesday book, compiled in 1086 in England with the purpose of acquainting William the Conqueror with his possessions.

The first published table of mortality, or life expectancy, is attributed to the Roman jurisconsul Emilius Mercer about the year 225 A.D. His table was considered quite inaccurate and Mercer authorized the publication of a more reliable table, credited to Ulpian. Ulpian's table was so accurate for its time that it was used for more than sixteen centuries [57].

Surprisingly, the first mathematical model of a population was not actually intended as a model but was simply a kind of entertainment problem. In the third section of his famous book *Liber abaci*, Leonardo Pisano—better known as Fibonacci—considered a problem of projecting a population of rabbits, starting from one pair and assuming that every month each pair produces another pair that in turn becomes reproductive in the second month. The solution to this problem led to the definition of a sequence of numbers known as the Fibonacci numbers. The problem fits into the Leslie-matrix framework that we shall describe later in this chapter.

A formula for the *expectancy of life*—that is, the average number of years a person of age x is expected to live—was proposed by Girolamo Cardano in 1570. He suggested that the expectancy of life is proportional to the number of years remaining to the maximal age. In other words, the expectancy of life is a linear decreasing function of age. This relationship was based on a single individual and was not generalized to populations. Thus, it has a limited applicability [116].

One of the most successful and, even today, widely used tools in demography is the set of basic life tables. The life tables follow the life of a group of people born together, called the *birth-cohort*, whose initial size is called the *radix*, usually 100,000. To compose a life table one has to record how many individuals from a birth-cohort survive to each birthday. From that information the probability of surviving from one age to another, the probability of dying within a certain age interval, the number of person-years lived by the cohort, and the life expectancy at birth can be computed and laid out in a table called the life table [57]. We show below in Table 1.1 the life table for the United States population in 1990 [122].

Table 1.1. *Life table by age, United States,* 1990.

x to $x+n$	${}_nq_x$	l_x	${}_nd_x$	${}_nL_x$	T_x	${}^\circ e_x$
0–1	0.00926	100,000	927	99,210	7,535,219	75.4
1–5	0.00184	99,073	183	395,863	7,436,009	75.1
5–10	0.00111	98,890	110	494,150	7,040,146	71.2
10–15	0.00128	98,780	127	493,654	6,545,996	66.3
15–20	0.00436	98,653	430	492,290	6,052,342	61.3
20–25	0.00549	98,223	539	489,794	5,560,052	56.6
25–30	0.00621	97,684	607	486,901	5,070,258	51.9
30–35	0.00765	97,077	743	483,571	4,583,357	47.2
35–40	0.00988	96,334	952	479,425	4,099,786	42.6
40–45	0.01261	95,382	1,203	474,117	3,620,361	38.0
45–50	0.01868	94,179	1,759	466,820	3,146,244	33.4
50–55	0.02905	92,420	2,685	455,809	2,679,424	29.0
55–60	0.04570	89,735	4,101	439,012	1.784,603	24.8
60–65	0.07058	85,634	6,044	413,879	7,535,219	20.8
65–70	0.10285	79,590	8,186	378,369	1,370,724	17.2
70–75	0.15191	71,404	10,847	330,846	992,355	13.9
75–80	0.22110	60,557	13,389	270,129	661,509	10.9
80–85	0.32387	47,168	15,276	197,857	391,380	8.3
85 and over	1.00000	31,892	31,892	193,523	193,523	6.1

The first column shows the age interval between the two exact ages indicated. The second column shows the proportion dying before reaching the end of that age interval. The third column lists the number surviving and constitutes the basis for all others. The fourth column shows the number dying in each successive age interval. The fifth and sixth columns show, respectively, the numbers of persons in the indicated age interval and in it and all subsequent age intervals for the stationary population. This is a population with exactly the mortality rate given in the second column and an annual birth rate of 100,000 distributed uniformly in time. The last column shows the average remaining lifetime at the end of each age interval—also called the expectation of life.

Actually, nowadays life tables are composed not only for the human population but also for nonhuman populations [28, 45].

The first important demographic work is considered to be John Graunt's *Natural and Political Observations Upon the Bills of Mortality* (from 1662), in which the author sketches the concept of a life table [116, 57]. Early contributions to the development of the life tables were also made by Edmund Halley (1693), Johan DeWit (1671), and Antoine Deparcieux

(1746) (as cited in [116]). Aside from the contribution of Daniel Bernoulli (1766) (as cited in [116]), who introduced continuous analysis and suggested the continuous age-structured force of mortality $\mu(x)$, and the models of Malthus and Verhulst that we discuss below, most of the work in mathematical demography before Lotka was directed at improving the life tables and expanding their use. An extensive overview of the history of the life tables can be found in [57, 116].

The life tables describe the life history of a birth-cohort of people with radix l_0. A sequence of such cohorts of size l_0 that are born every year, from which l_x individuals attain age x every year, forms a population with a constant size and fixed age composition. Such a population is called *stationary* [112]. A stationary population is closed to migration and has unchanging age-structured death rates, a constant number of births each year, and a zero growth rate. The basic life table is a mathematical model of a stationary population, that is, a population that does not change in size or age composition.

The continuous models of demographic growth are a natural generalization of life tables that explicitly include the dynamics of reproduction. The early models developed were age independent and, consequently, played a limited role in mathematical demography.

Thomas Malthus [81] is considered to have been the first to create a mathematical model of population growth in 1798. His model involves two demographic parameters, namely b, the per capita birth rate, and m, the per capita death rate. The Malthus model is based on the assumption that the rate of change of the total population is equal to the birth rate, given by bP, minus the death rate, given by mP, that is,

$$\frac{d}{dt}P(t) = bP(t) - mP(t).$$

Here t denotes time and $P(t)$ indicates the number of people present at time t. In other words, the rate of change of the total population is proportional to the population itself:

$$\frac{d}{dt}P(t) = rP(t),$$

where $r = b - m$. The solution of this model is exponential:

$$P(t) = P(0)e^{rt}.$$

Clearly, if $r = 0$ the population is stationary, if $r > 0$ the population grows exponentially without bound, and if $r < 0$ the population decays exponentially and approaches zero. Demographers today call populations that grow exponentially Malthusian populations, and their growth rate r is called the *Malthusian constant* or *intrinsic growth rate* of the population. Many models in demography are such that, asymptotically, their solutions grow exponentially.

Most continuous population models have discrete analogues that were developed either as a consequence or completely independently of a continuous model. In discrete models the population is counted at discrete times t that usually indicate the population size at successive generations, P_t. If R represents the number of individuals in the next generation per individual in the current generation, then the population at time $t + 1$ is given by

$$P_{t+1} = RP_t.$$

The solution of this equation can be found explicitly as the geometric sequence

$$P_t = R^t P_0,$$

which exhibits exponential growth. In fact, if $R > 1$ the population grows without bound, if $R = 1$ the population is stationary, and if $R < 1$ the population decreases to zero.

Exponential models cannot be accurate for long-term population projections because they predict either population explosion or population extinction. However, they can be used for projections on short time-scales. Their primary significance lies in the fact that the Malthusian parameter r summarizes the joint impact of the birth and death rates, with a clear implication about the general trend of the population: increasing or decreasing.

The main reason why natural populations cannot grow exponentially indefinitely is that their growth is regulated by various factors, which often include limitations in the food supply and available territory. Recognizing the inability of Malthus's model to capture the long-term population dynamics, Pierre-François Verhulst [123] proposed in 1838 that the per capita rate of population growth is not constant as in Malthus's model, but dependent on the population. He specifically suggested that the per capita growth rate is a linearly decreasing function of the population. This assumption leads to the following model, known as *Verhulst's model* or the *logistic equation*:

$$\dot{P}(t) = r\left(1 - \frac{P(t)}{K}\right) P(t),$$

where the constant K is called the *carrying capacity of the environment* and is equal to the population size at which the per capita growth rate of the population is zero. This equation has an explicit solution, $P(t) = Ke^{rt}/[(e^{rt}-1)+\frac{K}{P_0}]$. If the initial number of the population described by this equation is above the carrying capacity, then the population decreases in time to the value K; if the initial value of the population is below the carrying capacity, then the population increases nearly exponentially at the beginning, but then increases at a decreasing rate and tapers off, again approaching the carrying capacity—from below. The graph has the characteristic S-shape of a logistic curve, as shown in Figure 1.1.

Based on his theory Verhulst predicted that the carrying capacity for the population of Belgium would be 9.4 million people. The total population of Belgium as of January 2000 is 10.24 million people, a difference of only 0.84 million people—mostly due to immigration.

Just as Malthus's, the logistic equation also has a discrete analogue, the *logistic difference equation*:

$$P_{t+1} = (1+r)P_t - \frac{r}{K}P_t^2.$$

The behavior of the solutions of the discrete logistic equation is far more complicated than that of the solutions of the continuous logistic equation. In particular, for $0 < r < 1$ the populations described by the solutions tend monotonically to the carrying capacity K; for $1 < r < 2$ the solutions still tend to the carrying capacity but in oscillatory fashion; for r slightly larger than 2 there is a sequence of period-doublings that go from equilibrium to 2-cycle, 4-cycle, and so on. For $r > 2.57$ chaotic behavior can be detected [71]. An extensive account of density-dependent population equations of logistic type can be found in [2].

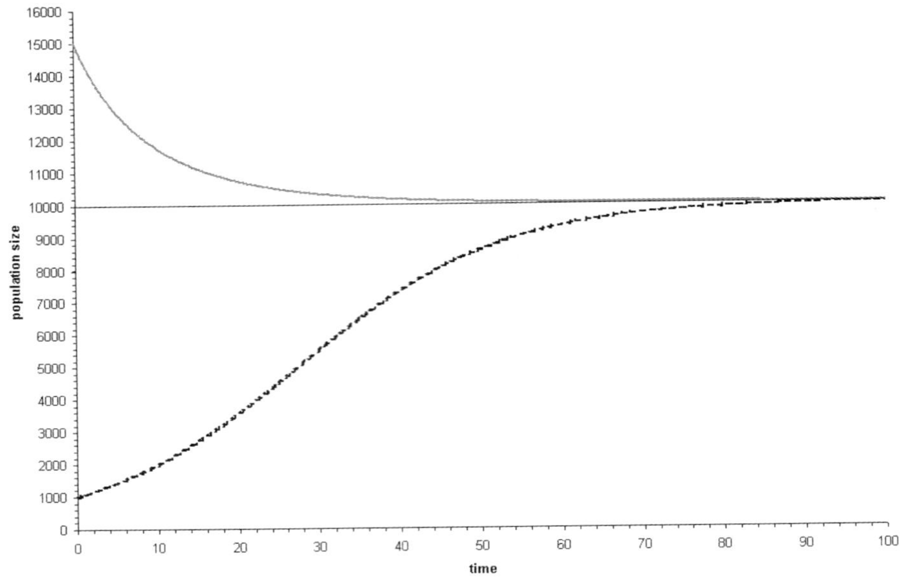

Figure 1.1. *Examples of logistic curves with carrying capacity* $K = 10{,}000$.

Among the mathematical models before Lotka's, perhaps another one that deserves mentioning is *Gompertz's law*. In 1825 Benjamin Gompertz [33] postulated that if $L(x)$ is the number of individuals living at age x, then it should vary according to

$$\frac{d}{dx}L = aq^x L \tag{1.1}$$

for appropriate constants a and q, $a < 0$ and $q > 1$. Thus, integrating this equation, one obtains that the number of individuals living at age x is given by a double exponential function called the *Gompertz curve*:

$$L(x) = cd^{q^x} \qquad (c = e^{-(a/\ln q)},\ d = 1/c).$$

Equation (1.1) was originally written as a population equation with a time variable instead of an age variable [71]. In that form it is called *Gompertz's law*. Gompertz's law was improved upon by William Makeham in articles from 1860 and 1867 [79, 80]. Makeham was also concerned with fitting the age-dependent mortality rate and the number of individuals still living at age x. He suggested the following formula:

$$L(x) = cs^x d^{q^x}.$$

In contrast to life tables that model stationary populations with *constant* numbers of births per unit of time, the population projection models of Malthus and Verhulst describe populations with birth rates that can vary in time. However, their major shortcoming is that they consider homogeneous populations and, in particular, they do not take into account the population age structure. Perhaps this is the main reason why their application in demography was limited.

1.2 The Lotka–McKendrick equation

The concept of "stable" population is attributed to Euler [30] who, in 1760, studied the problem of making projections for a population based on incomplete data. Euler considered a population with "stable" age structure—that is, time-invariant age-specific birth and death rates—and in which the proportion of each age group remains fixed.

The connection between the concept of a "stable" population and its mathematical formalization was made by Alfred Lotka (1907, 1922) in [76, 77] and F. R. Sharpe and Lotka (1911) in [114]. These three works are the backbone of modern demographic stable population theory.

Sharpe and Lotka's [114, 77] model describes the dynamics of the births and is based on an integral equation of Volterra type. In analyzing this model Lotka also realized that, though the age density of the population varies in time, it cannot take completely arbitrary values and it should eventually stabilize. Though he was not able to rigorously establish this fact, he offered geometric arguments [77] in support of the thesis.

The model is essentially a single-sex model derived under the assumption that the other sex has consistent behavior. We discuss it here because it forms the cornerstone of contemporary mathematical demography. Its connection to the stable population theory is discussed in section 1.3.

Let $B(t)$ be the density of female births at time t and let $u(x,t)$ be the age-specific density of females at time t, that is, $u(x,t)dx$ is the total number of women whose age is between x and $x+dx$. The number of women aged x to $x+dx$ at time t who were born since time zero is $B(t-x)dx$, and the proportion of those who have survived to age x (we assume here that $x \leq t$) is

$$B(t-x)\pi(x)dx,$$

where $\pi(x)$ is the probability of surviving from birth to age x. The number of women aged x to $x+dx$ at time t is related to the number of women aged $x-t$ to $x+dx-t$ at time 0, $u(x-t,0)\,dx$, by

$$\frac{\pi(x)}{\pi(x-t)}u(x-t,0)dx,$$

where $\frac{\pi(x)}{\pi(x-t)}$ gives the probability that a woman of age $(x-t)$ will survive to age x (here $x > t$). At time t all women aged x to $x+dx$ give birth to

$$\beta(x)u(x,t)dx$$

female children per unit time. The function $\beta(x)$ is the per capita rate of female childbearing among women aged x per unit time. Clearly, the ıtotal (female) birth rate at time t is

$$\begin{aligned} B(t) &= \int_0^{\omega} \beta(x)u(x,t)\,dx \\ &= \int_0^{t} \beta(x)B(t-x)\pi(x)dx + F(t), \end{aligned} \tag{1.2}$$

where

$$F(t) = \int_0^\omega \beta(x+t)u(x,0)\frac{\pi(x+t)}{\pi(x)}dx$$

and ω is the maximal possible age.

This is the basic equation of Sharpe and Lotka's deterministic one-sex population model. The function $K(x) = \beta(x)\pi(x)$ is called the *net maternity function*. It is related to the fundamental parameter

$$\mathcal{R}_0 = \int_0^\omega \beta(x)\pi(x)\,dx,$$

which gives the average number of children born to a female (male) during her (his) life. $\mathcal{R}_0$ is known as the *net reproduction rate* (NRR).

The NRR was introduced in demography by Richard Böckh [8] in 1886. It is clear from the physical interpretation of $\mathcal{R}_0$ that if $\mathcal{R}_0 > 1$ the population is increasing (that is, more people are being born than are dying), if $\mathcal{R}_0 < 1$ the population is decreasing (that is, more people are dying than are being born), and if $\mathcal{R}_0 = 1$ the population is stationary.

In terms of the maternity function, Sharpe and Lotka's equation can be written in the convolution form

$$B(t) = \int_0^t K(x)B(t-x)\,dx + F(t). \tag{1.3}$$

Equation (1.3) is often called the *renewal equation* or *Lotka's equation*. From a mathematical point of view it is a Volterra equation of the second kind and, under fairly general mathematical assumptions on the parameters of the model, a unique solution to the renewal equation always exists. A rigorous investigation of the relevant properties of the solution of the renewal equation can be found in [31].

This model can be developed by an alternative approach, first introduced in 1926 by A. C. McKendrick [89] and in 1959—in the context of cell growth modeling—by H. von Foerster [124]. This approach leads to a partial differential equation for the density function $u(x,t)$ introduced above, but the two models are continuous and essentially equivalent.

To introduce the McKendrick–von Foerster model we note that, if we consider a group of individuals with age x at time t, after a time Δt has elapsed, this same group will have age $x + \Delta t$ and the only change in the number of individuals is due to the fact that some of them died during the time interval $(t, t+\Delta t)$ (we assume that the population is closed). If the age-specific per capita mortality rate is $\mu(x)$, then

$$u(x+\Delta t, t+\Delta t) - u(x,t) = -\mu(x)u(x,t)\Delta t.$$

Dividing both sides by Δt and passing to the limit, we get

$$Du(x,t) = -\mu(x)u(x,t),$$

where $Du(x,t)$ is the directional derivative along the lines with slope one in the (x,t)-plane. Assuming that the partial derivatives exist, we obtain the model equation

$$\frac{\partial u}{\partial t} + \frac{\partial u}{\partial x} = -\mu(x)u(x,t).$$

If $\beta(x)$ is the age-specific per capita birth rate, then the *total birth rate* at the instant t is given by

$$B(t) = \int_0^\infty \beta(a)u(a,t)\,da$$

and is used as a boundary condition for this model. Hence, the model has the form

$$\begin{cases} \dfrac{\partial u}{\partial t} + \dfrac{\partial u}{\partial x} = -\mu(x)u(x,t), \\ u(0,t) = \displaystyle\int_0^\omega \beta(a)u(a,t)\,da, \\ u(x,0) = u_0(x), \end{cases} \tag{1.4}$$

where u_0 is a prescribed initial age density. The per capita birth rate $\beta(x)$ and the per capita death rate $\mu(x)$ are important characteristics of the population, usually called the *vital rates*.

As mentioned before, the McKendrick–von Foerster model is equivalent to the Sharpe–Lotka model. To see this, one can integrate the first equation of (1.4) along the characteristic lines, thus obtaining the age-specific density in terms of the total birth rate and the initial density:

$$u(x,t) = \begin{cases} u_0(x-t)\dfrac{\pi(x)}{\pi(x-t)} & \text{if } x \geq t, \\ B(t-x)\pi(x) & \text{if } t > x. \end{cases} \tag{1.5}$$

Then, one can substitute in the equation for the total birth rate, thus obtaining the Sharpe–Lotka integral equation (1.2). The quantity $\pi(x)$ is called the *survival probability* to age x, and in this case it is defined by the expression

$$\pi(x) = e^{-\int_0^x \mu(\tau)\,d\tau}. \tag{1.6}$$

The model (1.4) is applicable to either sex. When applied to just one of the sexes, it is implicitly assumed that the other sex has a development consistent with the one under consideration. Obviously, this approach cannot well capture situations in which one of the sexes is much scarcer than the other (this phenomenon is called *marriage squeeze*). The one-sex model is usually applied to females since they have a shorter reproductive span than males and births are more easily attributed to the mother than to the father.

1.3 The concept of stable population and the one-sex stable population theory

In investigating populations [22, 67] demographers make use of the concept of stable populations according to the following definition:

> *A* **stable population** *is a population that is closed to migration, in which the age-specific birth and death rates are time invariant.*

Some additional (but not essential) features are the assumptions that the population is one sex and there is some finite age beyond which no individual can survive.

A stable population is modeled by the McKendrick–von Foerster model (1.4). In a stable population the total birth rate increases (or decreases) exponentially in time at a constant rate λ (in contrast with a stationary population where the births are constant), that is,

$$B(t) = Ae^{\lambda t}.$$

A solution of the renewal equation of this form corresponds to a separable solution of (1.4) in the form $u(x,t) = e^{\lambda t}\mathcal{U}(x)$ (in fact it can easily be seen that all separable solutions of (1.4) are of this form). Looking for such solutions we substitute this expression in (1.4), thus obtaining for the age component of the separable solution the linear eigenvalue problem

$$\begin{cases} \mathcal{U}'(x) = -\lambda\mathcal{U}(x) - \mu(x)\mathcal{U}(x), \\ \mathcal{U}(0) = \displaystyle\int_0^\omega \beta(x)\mathcal{U}(x)\,dx. \end{cases}$$

Solving this initial-value problem for a first order ordinary differential equation, we obtain

$$\mathcal{U}(x) = \mathcal{U}(0)e^{-\lambda x}\pi(x), \tag{1.7}$$

where $\pi(x)$ is given by (1.6). If $\mathcal{U}(0)$ is equal to zero, we have a solution for every λ; otherwise the boundary condition can be satisfied if and only if the parameter λ satisfies the integral equation

$$\int_0^\omega e^{-\lambda x}\beta(x)\pi(x)\,dx = 1. \tag{1.8}$$

Equation (1.8) is called the *characteristic equation* for (1.4). It can be shown (a basic fact in mathematical demography [105]) that the characteristic equation (1.8) has a unique real solution λ^* and all its complex solutions λ occur in complex conjugate pairs with their real part smaller than the growth rate of the population:

$$\Re(\lambda) < \lambda^*.$$

The main result in stable population theory is based on the long-term behavior of the solutions of (1.4) (or the solutions of the renewal equation). It says that, given a female (male) age-specific mortality rate and a female (male) age-specific birth rate, then regardless of the initial age distribution, the population either vanishes asymptotically or converges asymptotically to a separable age distribution of the form $e^{\lambda^* t}\mathcal{U}(x)$, where $\mathcal{U}$ is called a *stable age distribution*. The fact that a population eventually "forgets" its initial age distribution is an important principle in demography called *strong ergodicity*.

It appears that the demographic term "stable age distribution" has its source not in its dynamical stability but rather in the fact that, for this solution, the proportion of each age group in the population remains constant in time:

$$\frac{u(x,t)}{\int_0^\omega u(x,t)\,dx} = \frac{\mathcal{U}(x)}{\int_0^\omega \mathcal{U}(x)\,dx}. \tag{1.9}$$

Mathematicians call solutions of the form $e^{\lambda^* t}\mathcal{U}(x)$ *persistent solutions* to distinguish them from equilibrium solutions, i.e., time-independent solutions, and from their dynamical stability or instability. The term "persistent solutions" is not completely justified (but is still used) in the case when $\lambda^* < 0$, since the population in this case does not persist but dies out.

The parameter λ^*—the unique real solution of the characteristic equation (1.8)—is an important demographic quantity called the *intrinsic (equilibrium) growth rate* of the population. However, to have an explicit formula for the exact value of the growth rate from given age-specific birth and death rates is in general not possible. Hence, this quantity has to be approximated. The first method for its computation was actually proposed by Lotka [78]. Lotka's approach is to expand the exponent in a Taylor series and represent the left-hand side of the equation as a power series with coefficients dependent on the moments of the maternity function. In particular, if the nth moment of the maternity function is given by

$$\mathcal{R}_n = \int_0^\infty x^n \beta(x)\pi(x)\,dx,$$

then

$$\int_0^\omega e^{-\lambda x}\beta(x)\pi(x)\,dx = \sum_{n=0}^\infty (-1)^n \frac{\mathcal{R}_n}{n!}\lambda^n.$$

Taking logarithms on both sides of the characteristic equation (1.8), expanding in a Taylor series again, and taking the first three terms of this series, one obtains a quadratic equation for the growth rate of the population:

$$\frac{1}{2}\kappa_2\lambda^2 - \kappa_1\lambda + \log\mathcal{R}_0 = 0,$$

where the coefficient κ_2 is the variance of the net maternity function given by

$$\kappa_2 = \frac{\mathcal{R}_1}{\mathcal{R}_0} - \left(\frac{\mathcal{R}_2}{\mathcal{R}_0}\right)^2.$$

The coefficient κ_1 is the mean of the net maternity function and is given by

$$\kappa_1 = \frac{\mathcal{R}_1}{\mathcal{R}_0}.$$

For realistic values this equation has two real solutions. The one that approximates the growth rate is the smaller one. Consequently,

$$\lambda^* = \frac{\kappa_1 - \sqrt{\kappa_1^2 - 2\kappa_2\log\mathcal{R}_0}}{\kappa_2}.$$

Lotka's method is usually described in terms of cumulants and a cumulant-generating function. An explicit form of this procedure was presented by Karmel [61]. Truncating the cumulant-generating function to the second cumulant represents in effect approximation of the net maternity function by a normal distribution.

Although Lotka's method is quite accurate, many other methods for the approximation of the intrinsic growth rate were created. An iterative method that is particularly well suited to the use of computers, as well as more accurate than Lotka's method, was proposed by Ansley Coale [20]. Coale makes no reference to Newton's iterations, but his method bears remarkable resemblance to Newton's method. Keyfitz also proposed an iterative procedure that was found to produce a solution accurate to six decimal places in four iterations [67]. Another iterative method that seems to have mostly theoretical significance was suggested by J. H. Pollard [105, 104]. This method is based on the use of the cumulant-generating function. James McCann devised a two-step method that eliminates the necessity to compute the second cumulant [86]. His method, however, proved less precise than Lotka's.

The intrinsic growth rate of the population is naturally related to the NRR. In fact $\mathcal{R}_0$ is the left-hand side of (1.8) when $\lambda = 0$. Looking at the definition of $\mathcal{R}_0$ and at (1.8) we can easily see that

$$\begin{array}{lll} \mathcal{R}_0 < 1 & \text{iff} & \lambda^* < 0, \\ \mathcal{R}_0 = 1 & \text{iff} & \lambda^* = 0, \\ \mathcal{R}_0 > 1 & \text{iff} & \lambda^* > 0. \end{array} \tag{1.10}$$

The convergence of the population density to the stable age distribution was rigorously proved for the first time by William Feller [31] and has been widely discussed since then. In [126] we find a proof that uses a semigroup technique. In [54] the author determines the limiting behavior of the total birth rate. The result established there reflects the concept that populations grow exponentially in an unlimited environment. The paper [10] (see also [54]) presents an alternative formulation of the McKendrick–von Foerster model. This formulation is in terms of the per capita age density called the *age profile*. In particular, if the total population size is denoted by $P(t)$ and is represented by the integral of the age density, then the proportion of individuals of age x in the total population—called the age profile—is denoted by $\mathcal{U}(x, t)$:

$$\mathcal{U}(x,t) = \frac{u(x,t)}{P(t)}. \tag{1.11}$$

By integrating the age density u over all ages we can see that the integral of the age profile must be equal to one for all t. Thus, the age profile can be viewed as a probability density function. We can recover the age density as the product of the total population size and the age profile. Differentiating the formula for the total population size and using (1.4) we obtain a Malthus-type differential equation for the total population:

$$\frac{d}{dt}P(t) = \alpha(t)P(t),$$

where the per capita rate of change α depends on time through the age profile

$$\alpha(t) = \int_0^{\omega} [\beta(x) - \mu(x)]\mathcal{U}(x,t)\,dx.$$

This equation for the total population uncovers another relationship between the McKendrick–von Foerster model (1.4) and Malthus's—the former model is a generalization of the latter because it also allows for time-variable growth rates.

Differentiating formula (1.11) in age and time, we obtain the following initial boundary value problem for the profile:

$$\begin{cases} \mathcal{U}_t(x,t) + \mathcal{U}_x(x,t) = -\mu(x)\mathcal{U}(x,t) - \alpha(t)\mathcal{U}(x,t), \\ \mathcal{U}(0,t) = \int_0^\omega \beta(x)\mathcal{U}(x,t)\,dx, \\ \int_0^\omega \mathcal{U}(x,t)\,dx = 1, \\ \mathcal{U}(x,0) = \mathcal{U}_0(x). \end{cases} \tag{1.12}$$

Problem (1.12) is closed and can be solved by itself. It is established in [54] that it has a unique solution. After the age profile is obtained, the equation for the total population size can be solved since it depends on the age profile through $\alpha(t)$:

$$P(t) = P(0)e^{\int_0^t \alpha(s)ds}.$$

In contrast with the McKendrick–von Foerster model (1.4), which is linear, the differential equation for the profile is nonlinear because of the term $\alpha(t)\mathcal{U}$. This equation has a time-independent solution:

$$\mathcal{U}^*(x) = \frac{e^{-\lambda^* x}\pi(x)}{\int_0^\infty e^{-\lambda^* x}\pi(x)\,dx},$$

where the intrinsic growth rate of the population λ^* is determined as the unique real solution of the characteristic equation (1.8). This is the same expression as (1.7) with $\mathcal{U}(0) = (\int_0^\infty e^{-\lambda x}\pi(x)dx)^{-1}$. It is proved in [54] that the age profile $\mathcal{U}(x,t)$ approaches the equilibrium age profile $\mathcal{U}^*(x)$ as time goes to infinity for almost all values of x. Consequently, as time goes to infinity, $\alpha(t)$ approaches λ^*.

The considerations on the age profile explain why the proportion of each age group in the total population remains constant in time for stable populations. In fact this proportion—the age profile—is not constant in time in general, but it approaches the ultimate age profile $\mathcal{U}^*(x)$ as time goes to infinity. Hence, this proportion can be approximated by $\mathcal{U}^*$.

1.4 Leslie's discrete one-sex stable population theory

Lotka and Sharpe's and Lotka's models are continuous. It is necessary that the vital rates be given as continuous functions for these models to be applied. However, these are usually collected as empirical data and are readily available in discrete form, so they can be more easily used in a discrete model. The first work to use matrix algebra for population projection is by Harro Bernardelli, and it appeared in 1941 [5]. Bernardelli was concerned with the oscillations of human population dynamics, rather than with its long-term stabilization. Matrix population models were also suggested by E. G. Lewis in a paper that appeared in 1942 [74]. Lewis essentially suggests the first discrete model and considers the projection matrix associated with it and its characteristic equation. Patrick Holt Leslie, who was apparently unaware of Lewis's article, rediscovered the model in his 1945 paper [73]. Leslie considered the mathematical properties of the model more thoroughly as well as the resulting convergence to a stable age distribution. Although Leslie presented the matrix approach to stable population theory, his results were ignored by both ecologists and demographers for the following twenty years. Interest in Leslie models in demography was aroused after Keyfitz [66] and Goodman [35] demonstrated their equivalence to the existing continuous theory.

In the remainder of this section we introduce Leslie's matrix method [73]. As with Lotka's model, only one-sex populations are taken into account. Time is considered at discrete intervals of specified length. Age is also broken into age groups that correspond to the unit interval of time.

Let $n_{x,t}$ be the number of individuals in the age group x to $x+1$ at time t. We assume there are $m+1$ age groups. The probability that an individual aged x to $x+1$ at time t will survive to be in the age group $x+1$ to $x+2$ at time $t+1$ is given by P_x, called the *year-to-year survival probability*. The probability P_x is strictly positive for $x < m$ and we have the relations

$$\begin{cases} n_{1,t+1} = P_0 n_{0,t}, \\ n_{2,t+1} = P_1 n_{1,t}, \\ \vdots \\ n_{m,t+1} = P_{m-1} n_{m-1,t}. \end{cases} \tag{1.13}$$

It is assumed that $P_m = 0$—that is, nobody survives beyond age m. The survival probability from one age group to the next can be computed as the ratio of the probability of survival to age $x+1$ and the probability of survival to age x:

$$P_x = \frac{\pi_{x+1}}{\pi_x},$$

where π_x is the probability of survival from birth to age x and can be taken from the life tables. Furthermore, if the number of progeny born per individual aged x to $x+1$ in the time interval t to $t+1$ is given by β_x, the *fertility* function for this age group is given by

$$F_x = P_x \beta_{x+1}. \tag{1.14}$$

The fertility function gives the number of offspring produced by an individual aged x to $x+1$ reduced by the survival probability for that age class. It is the discrete analogue of the maternity function. Thus, the total number of newborns during the period t to $t+1$ is given by the relation

$$n_{0,t+1} = \sum_{x=0}^{m} F_x n_{x,t}. \tag{1.15}$$

Some of the fertilities F_x may be zero. In fact the ages x for which $F_x \neq 0$ constitute the so-called fertility window.

Leslie's model can also be written in matrix notation. If $\vec{\mathbf{n}}_t \equiv (n_{0,t}, \ldots, n_{m,t})$ is the vector of population numbers at time t, then

$$\vec{\mathbf{n}}_{t+1} = \mathbf{M}\vec{\mathbf{n}}_t, \tag{1.16}$$

where **M** is called a *Leslie matrix* or a *projection matrix*. It is made up of the coefficients of the model:

$$\mathbf{M} = \begin{pmatrix} F_0 & F_1 & \cdots & F_{m-1} & F_m \\ P_0 & 0 & \cdots & 0 & 0 \\ 0 & P_1 & \cdots & 0 & 0 \\ & & \cdots & & \\ 0 & 0 & \cdots & P_{m-1} & 0 \end{pmatrix}. \qquad (1.17)$$

The matrix is a square of dimension $(m+1) \times (m+1)$. If $F_m = 0$, then it is singular since the last column consists only of zeros. From (1.16) we see that the population at time t is given in terms of the initial population by the formula

$$\vec{\mathbf{n}}_t = \mathbf{M}^t \vec{\mathbf{n}}_0.$$

Within this discrete context the persistent solutions we introduced in the previous section take the form

$$\vec{\mathbf{n}}_t = \lambda^t \vec{\mathbf{v}},$$

where $\vec{\mathbf{v}} \equiv (v_0, \ldots, v_m)$ is a stable age distribution. Substituting this expression into (1.16) we find that the stable age distribution $\vec{\mathbf{v}}$ and λ must be solutions of the eigenvalue problem

$$\mathbf{M}\vec{\mathbf{v}} = \lambda \vec{\mathbf{v}}.$$

This equation has a nonzero solution if and only if λ is an eigenvalue of **M**. Since the matrix **M** has a very special form, to derive the characteristic equation we write the eigenvalue problem explicitly and solve iteratively. In particular, we have

$$v_1 = \frac{P_0}{\lambda} v_0,$$

and, recursively, we obtain

$$v_i = \frac{P_0 P_1 \cdots P_{i-1}}{\lambda^i} v_0, \quad i = 1, \ldots, m.$$

Substituting into the equation for the newborns one obtains the characteristic equation

$$F_0 \lambda^{-1} + P_0 F_1 \lambda^{-2} + \cdots + P_0 P_1 \cdots P_{m-2} F_{m-1} \lambda^{-m} = 1. \qquad (1.18)$$

Using the representation of the fertility (1.14) and noticing that

$$\pi_x = P_0 P_1 \cdots P_{x-1}$$

we can rewrite the characteristic equation as

$$\sum_{x=1}^{m} \beta_x \pi_x \lambda^{-x} = 1,$$

a discrete analogue of (1.8) with $\ln \lambda$ replacing λ. This is the discrete Lotka equation, which has exactly one positive real root λ^*. The remaining roots are either negative or complex and have modulus less than or equal to λ^*. More precisely, if the greatest common divisor p of the nonzero fertility entries of the Leslie matrix is one, then it is said that the fertility schedule is *aperiodic* and

$$|\lambda_j| < \lambda^*$$

for any root $\lambda_j \neq \lambda^*$ of the characteristic equation. If the fertility schedule is not aperiodic, it is said to be *periodic* with period p and then there are $p-1$ other roots of the characteristic equation that have the same modulus as λ^*. The solutions of the Leslie model asymptotically approach exponential growth given by

$$\vec{\mathbf{n}}_t \cong c_0(\lambda^*)^t \vec{\mathbf{v}}^*,$$

where $\vec{\mathbf{v}}^*$ is a unit eigenvector corresponding to the dominant eigenvalue λ^*. The stable age distribution of the population is given by

$$n_{x,t} \cong c_0 \pi_x (\lambda^*)^{t-x}.$$

1.5 The Gurtin–MacCamy equation

Lotka's and Leslie's models cannot be used for long-term projections since, just as Malthus's model, they predict population explosion or extinction. The main reason for this type of long-term behavior is that all these models are represented by linear equations. In contrast, Verhulst's model—created to remedy the inability of Malthus's to capture the long-term behavior of natural populations—is nonlinear. It is generated by assuming that the growth rate of the population depends on the total population. In particular, this means that the per capita birth rate and the per capita death rate are assumed to depend on the total population. Following this idea, Morton Gurtin and Richard MacCamy [37] modified the McKendrick–von Foerster model so that the per capita age-structured birth rate and the per capita age-structured death rate depend on the total population. To obtain their model assume that the birth rate $\beta(x, P)$ and the death rate $\mu(x, P)$ in (1.4) depend on the total population P. If $u(x, t)$ is the age density of individuals in the population, then it satisfies the initial boundary value problem

$$\begin{aligned} &\frac{\partial u}{\partial t} + \frac{\partial u}{\partial x} = -\mu(x, P(t))u(x, t), \\ &u(0, t) = \int_0^\omega \beta(x, P(t))u(x, t)\,dx, \\ &u(x, 0) = u_0(x), \end{aligned} \tag{1.19}$$

and the total population size is given by the integral of the age density

$$P(t) = \int_0^\omega u(x, t)\,dx.$$

As with the McKendrick–von Foerster model, the Gurtin–MacCamy model can also be integrated along the charateristic lines, a procedure described in [37], to obtain an implicit form of the solution:

$$u(x,t) = \begin{cases} u_0(x-t)\pi(t;\,P) & \text{if} \quad x \geq t, \\ \\ B(t-x)\pi(x;\,P) & \text{if} \quad t > x. \end{cases} \tag{1.20}$$

The quantity $\pi(z;\,P)$ is called the *density-dependent survival probability* and in this case is defined by the expression

$$\pi(z;\,P) = e^{-\int_0^z \mu(x-\tau, P(t-\tau))\,d\tau}.$$

The total birth rate is again given by $B(t) = u(0,t)$. If the implicit solution (1.20) is substituted into the equation for the total birth rate, an analogue of the Sharpe–Lotka integral equation (1.2) is obtained:

$$B(t) = \int_0^t \beta(x, P(t))\pi(x;\,P)B(t-x)\,dx + F_1(t).$$

Unlike the linear case, the equation for the total birth rate here is not closed because it depends on the total population. Thus, it has to be coupled with an integral equation for the total population size:

$$P(t) = \int_0^t \pi(x;\,P)B(t-x)\,dx + F_2(t).$$

The system of integral equations for the total birth rate and the total population size represents the nonlinear analogue of the Sharpe–Lotka model.

Gurtin and MacCamy show that the model is well posed, which means that for any initial age distribution $u_0(x)$ there is a unique nonnegative solution of the model (1.19). The only requirement for this to hold is that the age-specific birth rate is bounded above by a constant $\bar{\beta}$. If $\underline{\mu}$ is the infimum of the death rate, then they establish that the total birth rate and the total population size grow exponentially at most with growth rate $\bar{\beta} - \underline{\mu}$, which they call the *bounding growth rate*. Despite this fact, the density-dependent model has a very different long-term behavior than the linear model. Gurtin and MacCamy discuss conditions in order the total population converges, as time goes to infinity, to a finite constant value P (a local result) and the age density converges to a time-independent age distribution $v(x)$. The limit values satisfy the equations

$$\begin{aligned} v'(x) &= -\mu(x, P)v(x), \\ v(0) &= \int_0^\omega \beta(x, P)v(x)\,dx, \end{aligned} \tag{1.21}$$

where

$$P = \int_0^\omega v(x)\,dx.$$

The solution $v(x)$ of this problem is called an *equilibrium age distribution*. The probability that a person will survive to age x when the population is held constant at level P is given in this case by

$$\pi(x;\,P) = e^{\int_0^x \mu(a,P)\,da}.$$

The NRR of a density-dependent population depends on the total population and gives the number of children born to an individual during her/his lifetime if the total population is held constant at P:

$$\mathcal{R}(P) = \int_0^\omega \beta(x, P)\pi(x; P)\,dx.$$

If the differential equation in (1.21) is solved and $v(x)$ is substituted into the initial condition, it follows that the equilibrium total population size must satisfy the equation

$$\mathcal{R}(P) = 1. \tag{1.22}$$

Once P is determined, the unique age distribution that corresponds to it is given by

$$v(x) = \frac{P\pi(x; P)}{\int_0^\omega \pi(x; P)\,dx}.$$

System (1.21) always admits the trivial solution $v \equiv 0$. In addition, (1.22) may have no nontrivial solutions, exactly one nontrivial solution, or more than one nontrivial solution. The behavior of the function $\mathcal{R}(P)$ depends on the mechanism of growth. It is realistic to expect that when P is small, $\mathcal{R}(P)$ increases as a function of P. This behavior is called the *Allee effect*. Then, there is some critical value P_0 so that $\mathcal{R}(P)$ decreases as a function of P for $P > P_0$ and tends to zero as P grows without bound. The decrease of the NRR for large population numbers is called the *logistic effect*. In this case if $\mathcal{R}(0) < 1$, then (1.22) may have two nontrivial solutions or no nontrivial solutions, depending on whether the maximum of the function $\mathcal{R}(P)$ is larger or smaller than one. If $\mathcal{R}(0) > 1$, then (1.22) always has exactly one nontrivial solution. If the function $\mathcal{R}(P)$ is strictly decreasing for $P \geq 0$ and approaches zero as P goes to infinity, then there is exactly one nontrivial solution if and only if $\mathcal{R}(0) > 1$. This situation is called a *purely logistic case*.

To illustrate the behavior described above we consider a specific example largely borrowed from [54], which can be consulted for the details. Assume the death rate depends only on age but not on the total population size: $\mu(x, P) = \mu_0(x)$. The corresponding probability of survival is denoted by $\pi_0(x)$. Next, assume that the age-specific birth rate is *separable*—that is, it consists of the product of a function that depends only on age, $\beta_0(x)$, and a function of the population size:

$$\beta(x, P) = \beta_0(x)e^{\epsilon P(1-\frac{P}{K})},$$

where K and ϵ are parameters. Then, the population-dependent NRR is

$$\mathcal{R}(P) = \mathcal{R}_0 e^{\epsilon P(1-\frac{P}{K})},$$

where

$$\mathcal{R}_0 = \int_0^\omega \beta_0(x)\pi_0(x)\,dx$$

is the *intrinsic* NRR. The dependence of the NRR on P is exemplified in Figure 1.2.

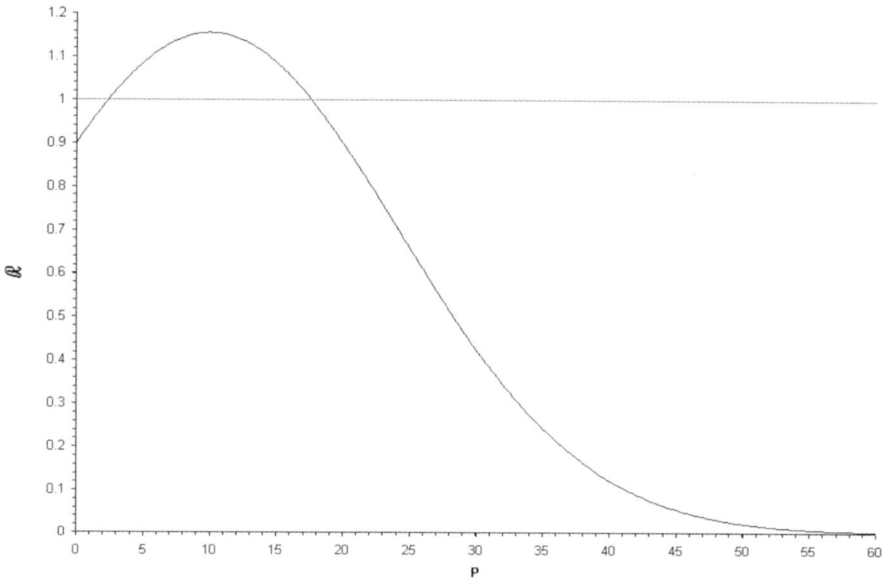

Figure 1.2. NRR $\mathcal{R}(P) = 0.9\, e^{0.05P(1-P/20)}$ ($P_0 = 10.0$).

Thus, the equilibrium population size must satisfy

$$\epsilon P \left(1 - \frac{P}{K}\right) = -\ln \mathcal{R}_0.$$

This equation has one nontrivial solution if $\mathcal{R}_0 > 1$, two nontrivial solutions if

$$e^{-\frac{\epsilon K}{4}} < \mathcal{R}_0 < 1,$$

and no nontrivial solutions if

$$0 < \mathcal{R}_0 < e^{-\frac{\epsilon K}{4}}.$$

If the equilibrium population size is plotted against the reproduction number $\mathcal{R}_0$, the curve bifurcates from the critical value $\mathcal{R}_0 = 1$ not forward but backward. Hence, for values of $\mathcal{R}_0$ slightly below 1 there are two nontrivial equilibria (see Figure 1.3). In this case it is said that the model exhibits a *backward bifurcation*.

Typically, the smaller equilibrium value is unstable while the larger equilibrium value is locally stable. Thus, depending on the initial values, the total population size may tend in time either to zero or to the larger nontrivial equilibrium value. In other words, the long-term behavior of the solution depends on the initial values, i.e., the model is *not ergodic*.

This example elucidates another behavior that is typical only for the nonlinear model. While for the McKendrick–von Foerster model the population persists and grows exponentially when $\mathcal{R}_0 > 1$ and dies out when $\mathcal{R}_0 < 1$, with the nonlinear model it is possible for

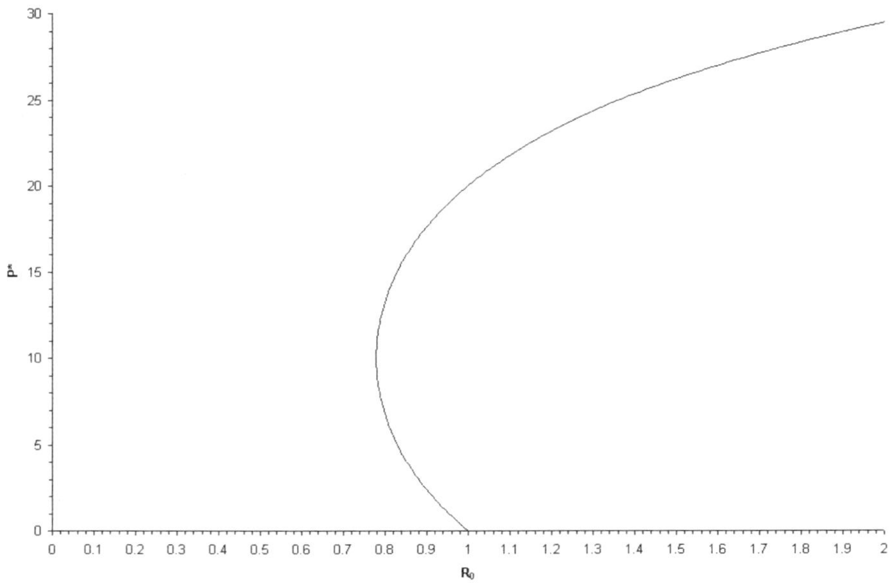

Figure 1.3. *Backward bifurcation for* $\mathcal{R}(P) = 0.9\, e^{0.05P(1-P/20)}$.

the population to persist and remain nonzero in the long run even when the intrinsic NRR $\mathcal{R}_0$ is smaller than one.

The previous nonlinear models can also be treated within the alternative framework of semigroup theory as done, for example, in [125].

1.6 Significance of the stable population theory and its extensions

Many of the assumptions on stable populations are not valid for natural populations. Nonetheless, the stable population theory often turns out to be a close approximation of reality. It permits demographers to investigate such basic characteristics of populations as the intrinsic growth rate and the ultimate stable age distribution. Although extended projections with the McKendrick–von Foerster model are not accurate, they help demographers better understand the long-term implications of the present vital rates. The stable theory is particularly helpful in the case of inconsistent or incomplete birth and death data. Furthermore, the one-sex population theory enables demographers to approach such problems as estimating how a given level of immigration will change the ultimate stable population or how birth control applied to women of various ages will influence the rate of increase of the population; tracing back the higher mean age of a country to low birth rates, rather than to a lower mortality [70]; and demonstrating that decreasing mortality does not necessarily make the population older [68].

The stable population theory develops a machinery for the investigation of the long-term behavior of the population. However, the models of the stable theory are built and valid under a set of assumptions that in reality are frequently not true. As a central concept in demography, the stable population theory has always attracted the interest of scientists. Various extensions are known today that remove one or more of the restrictions of the classical theory. One of the first extensions discussed was allowing the birth and death rates to depend on time. In fact, improvement of living conditions and advancement of medical science and technology have led to decreasing death rates and increasing life expectancy in the last four to five centuries. A model with time-dependent vital rates is known to have first been considered by Langhaar [72] (as cited in [54]) and investigated more thoroughly by Inaba [58]. Gurtin and McCamy [37] consider and analyze a nonlinear model whose vital rates depend on the total population. We pointed out in the previous section that this model does not exhibit exponential growth. Sinestrari [115] considers a nonlinear model with vital rates dependent on the age-specific density of the population. Models involving migration and diffusion have also been considered.

One of the most natural extensions of the one-sex theory is to include both sexes. The one-sex stable population theory is a fundamental theoretical tool in mathematical demography, but it cannot answer questions related to the existence and interplay of the sexes. When applied to one of the sexes, it is implicitly assumed that the other sex has a consistent development, but this approach cannot capture well situations in which one of the sexes is scarcer than the other. In human populations such a disproportion can be observed when the so-called marriage squeeze occurs [112]. Thus, a fundamental shortcoming of the classical stable population theory is that it fails to handle populations differentiated by sex.

The extension of the stable population theory to two-sex populations is, however, not trivial. Applying the one-sex model to both sexes simultaneously leads in most cases to contradictory results. In 1932 R. Kuczynski calculated the male and female net reproductive rates for France for the year 1920 and found the male rate to be 1.194 and the female rate to be 0.977 (as cited in [105]). Hence, if a one-sex all-female model had been used for France at the time, it would have projected a continuously decreasing population, while if a one-sex all-male model had been used, it would have predicted a continuously increasing population. In that case, even if only one of the one-sex models were used, it would certainly have given incomplete and partial information about the population of France. When one calculates the equilibrium growth rates for females and males separately, they are also likely to differ. This means that, in the long run, the sex ratio—i.e., the ratio of the density of the males to the density of the females—tends either to zero or to infinity.

The problem of eliminating inconsistencies inherent in one-sex population models and reconciling the female and male growth rates has been central in demography for many years. The next section presents early attempts to produce models suitable in this respect.

1.7 Early attempts to incorporate both sexes into one-sex population models

The reconciliation of the female and male growth rates has been addressed in demography repeatedly since the late 1940s. After Kuczynski first reported the problem of the inconsistency of these growth rates (as we mentioned in the previous section), P. H. Karmel reinforced it by analyzing the sources and magnitudes of the discrepancy between the female and male NRRs [62, 63]. Furthermore, Karmel made one of the earliest attempts to correct this problem. In [61] he begins with the assumption that the male and female growth rates must be equal and works to reconcile the stable age distributions. Apparently, he was the first to realize that a consistent model should be possible if nuptiality is introduced. Karmel believed that "... this inconsistency between the male and female age-specific fertility rates can be overcome in a population where monogamous marriage is the rule...." He also observed that, although the introduction of nuptiality eliminates the inconsistencies in the birth rates, it does so while introducing inconsistencies in the joint-nuptiality rates. These inconsistencies are expressed in the fact that the number of marriages between brides and grooms of given ages is not the same when projected on the basis of the female marriage probabilities or on the basis of the male marriage probabilities. Furthermore, Karmel noticed that it is impossible to hold both nuptiality rates constant: in fact they have to vary as the population changes to assume its stable distribution. This idea was further developed by J. Hajnal [44], who effectively suggested, without formalizing it, the necessity of a function which—based on the present marriage habits and given population numbers—can describe future nuptiality. In his answer to Hajnal's comments Karmel expressed the idea that the two-sex growth rate has to lie between the female and male growth rates [64].

Although many of the ideas in two-sex population modeling were first expressed in Karmel's papers, the first significant dynamical models were presented by A. H. Pollard [103] and D. G. Kendall [65]. Pollard uses integral equations to model the male, female, and total birth rates. He considers the female births to males and male births to females. In particular, if $\pi_f(x)$ and $\pi_m(y)$ are, respectively, the female and male probabilities of survival (see (1.6)), and $\beta_{mf}(x)$ and $\beta_{fm}(y)$ are, respectively, the male birth rate to females and female birth rate to males, then for $t > \omega$ we have

$$\begin{aligned} F(t) &= \int_0^\omega \beta_{fm}(y)\pi_m(y)M(t-y)dy, \\ M(t) &= \int_0^\omega \beta_{mf}(x)\pi_f(x)F(t-x)dx, \end{aligned} \tag{1.23}$$

where $F(t)$ and $M(t)$ are, respectively, the female and male total birth rates. Substituting the total male birth rate from the second equation into the first equation and vice versa we obtain the following self-contained equations for the total female and male birth rates:

$$\begin{aligned} F(t) &= \int_0^\omega \int_0^\omega \beta_{fm}(y)\beta_{mf}(x)\pi_f(x)\pi_m(y)F(t-x-y)dxdy, \\ M(t) &= \int_0^\omega \int_0^\omega \beta_{fm}(y)\beta_{mf}(x)\pi_f(x)\pi_m(y)M(t-x-y)dxdy. \end{aligned} \tag{1.24}$$

The total birth rate $B(t) = F(t) + M(t)$ satisfies the equation

$$B(t) = \int_0^\omega \int_0^\omega \beta_{fm}(y)\beta_{mf}(x)\pi_f(x)\pi_m(y)B(t-x-y)dxdy.$$

This is the simplest age-structured two-sex model that has the same growth rate for females and males. Looking for solutions of the form $B(t) = Be^{rt}$ one obtains the two-sex characteristic equation

$$\int_0^{\omega}\int_0^{\omega} e^{-(x+y)r}\beta_{fm}(y)\beta_{mf}(x)\pi_f(x)\pi_m(y)dxdy = 1.$$

This equation has a unique real solution r^*, and all complex solutions occur in conjugate pairs and have a real part smaller than the real solution. It can be shown that the growth rate computed from this model does indeed lie between the female and male growth rates.

Although A. H. Pollard's model is consistent and has the desired properties, it is very artificial since it is difficult to explain why one should consider female births to males and vice versa. In addition, this model is linear, while a fundamental expectation is that two-sex models should be nonlinear because they incorporate the interaction between the two sexes. A discrete version of this model was also offered by J. H. Pollard [105].

The two-sex models known and used today are based on two models by Kendall that appeared in 1949 [65]. Kendall's models do not involve age structure but are intrinsically nonlinear. The first model describes the dynamics of the numbers of females and males in the population. The nonlinear term is the total number of births per unit of time. The model is given by the system

$$\begin{aligned} F' &= -\mu F + \frac{1}{2}\Lambda(F, M), \\ M' &= -\mu M + \frac{1}{2}\Lambda(F, M), \end{aligned} \tag{1.25}$$

where $F(t)$ and $M(t)$ are, respectively, the numbers of females and males in the population. The function $\Lambda(F, M)$ is the total birth rate and μ is the per capita death rate. It is assumed that half of the births are girls and half are boys. Subtracting the two differential equations and integrating in time from 0 to t we obtain

$$M(t) - F(t) = (M(0) - F(0))e^{-\mu t},$$

which implies that any initial difference in the numbers of females and males will disappear in time. Moreover, the number of females remains dominant for all time if and only if it dominates at time zero. Kendall observed that if the birth rate is taken to represent random mating, $\Lambda(F, M) = FM$, then the model predicts population explosion in finite time. Thus, he concluded that in order to avoid difficulties, the birth rate should be linear in the total population size—in other words, it should be homogeneous of degree one.

The other model considered by Kendall is one that discriminates between married and unmarried individuals and involves marriages. If $F(t)$, $M(t)$, and $C(t)$ are, respectively, the numbers of single females, single males, and couples, then the dynamics of these three classes is given by the equations

$$\begin{cases} M' = -\mu M + \beta C + \mu C - \mathcal{M}(M, F), \\ F' = -\mu F + \beta C + \mu C - \mathcal{M}(M, F), \\ C' = -2\mu C + \mathcal{M}(M, F), \end{cases} \tag{1.26}$$

where β is the per couple birth rate and $\mathcal{M}(M, F)$ is the marriage rate, also called the *marriage function*. As before, it can be seen that the difference in the numbers of single females and single males tends to zero in time, keeping a constant sign. Naturally, the behavior of the solutions depends on the form assumed for the marriage function. Kendall himself considered only one specific marriage function,

$$\mathcal{M}(M, F) = 2\nu \min(M, F).$$

In this case the model reduces to a linear one dependent on the initial conditions. The solutions of that linear model are of the form $Ae^{p_1 t} + Be^{p_2 t}$, where p_1 and p_2 are the roots of the quadratic equation

$$p^2 + (3\mu + 2\nu)p + 2\mu^2 + 2\nu(\mu - \beta) = 0.$$

It can be seen that the population tends to infinity, approaches a finite limit, or tends to zero depending on whether $\beta > \lambda_1$, $\beta = \lambda_1$, or $\beta < \lambda_1$, where

$$\lambda_1 = \mu\left(1 + \frac{\mu}{\nu}\right).$$

Kendall's models inspired a lot of work on two-sex populations. L. A. Goodman [34] noticed that the assumption of equal birth and death rates for males and females can be weakened. He considered the same models but with different rates and several marriage functions, and he computed the ultimate sex ratio in many cases. Goodman also created simple discrete age–sex structured models [36]. The births were attributed to the females.

J. H. Pollard considered Kendall's model with couples with a number of different marriage functions [105]. Through simulations with the arithmetic mean marriage function

$$\mathcal{M}(F, M) = \frac{1}{2}(F + M),$$

Pollard finds that the model may lose its validity because it is possible for the number of single females to become negative. This is a consequence of the chosen marriage function allowing more marriages than there are singles, since

$$\min\{F, M\} < \frac{1}{2}(F + M) \text{ if } F \neq M.$$

Chapter 2

Gender Structure and the Problem of Modeling Marriages

Falstaff: No quips now, Pistol. Indeed, I am in the waist two yards about; but I am now about no waste; I am about thrift. Briefly, I do mean to make love to Ford's wife: I spy entertainment in her; she discourses, she carves, she gives the leer of invitation: I can construe the action of her familiar style; and the hardest voice of her behaviour, to be Englished rightly, is, "I am Sir John Falstaff's."

Falstaff: I have writ me here a letter to her; and here another to Page's wife, who even now gave me good eyes too, examined my parts with most judicious oeillades: sometimes the beam of her view gilded my foot, sometimes my portly belly.

—W. Shakespeare, *The Merry Wives of Windsor,*
Act I, Scene III, 1597

"Bokanovsky's Process," repeated the Director, and the students underlined the words in their little notebooks.

One egg, one embryo, one adult—normality. But a bokanovskified egg will bud, will proliferate, will divide. From eight to ninety-six buds, and every bud will grow into a perfectly formed embryo, and every embryo into a full-sized adult. Making ninety-six human beings grow where only one grew before. Progress.

—A. Huxley, *Brave New World,* Chapter 1, 1932

We all know that Falstaff did not succeed in his attempt to break the marriages of the Pages and Fords and we know that Anne Page and Fenton eventually married and (we hope) lived happily ever after. Actually, ever since, the game of couple formation, together with a fertility table, has been the basic mechanism for describing human reproduction.

Nowadays, although the temptations of a brave new world knock on the door, the good old mechanism is still in force.

This chapter is devoted to the presentation and setup of a basic demographic model built on the ideas formulated in those early results by Kendall that we mentioned at the end of the previous chapter. In the first section we shall discuss the early attempts to model the two-sex interaction based on the models by Kendall. Then, in section 2.2, we give a detailed presentation of the Fredrickson–Hoppensteadt model that we intend to adopt in the subsequent chapters. This model takes into account age structure and couple formation and it appears to provide a good framework for the investigation of the dynamics and growth of human populations. We also consider in sections 2.3 and 2.4 some significant variations of the model that may also be considered extensions of it. The second half of the chapter is focused on the marriage function, which is the main ingredient of the Fredrickson–Hoppensteadt model, and we discuss it from both the theoretical side and the empirical. On this subject we try to present different aspects and different approaches.

2.1 Historic overview of discrete and continuous two-sex population models

Kendall's models, presented in section 1.7, attracted the interest of scientists because their description of the reproduction dynamics seemed to be more adequate for an efficient model of the growth of a population. Thus, gender structure began to occupy a significant place in mathematical demography. Attention is now focused on the reproduction rate as a function of male and female densities. In addition, the *marriage function* plays a central role since in the second of Kendall's models the reproduction mechanism is mediated by the formation of couples.

Yellin and Samuelson [127] presented an analysis of the second of Kendall's models with a general form of the marriage function. It appears that they were the first to formalize the condition that the marriage rate be *homogeneous of degree one*:

$$\mathcal{M}(\alpha s_f, \alpha s_m) = \alpha \mathcal{M}(s_f, s_m),$$

where $\alpha > 0$ is a constant and s_f and s_m are, respectively, the densities of single females and males. Their paper gives conditions sufficient to ensure the existence of a unique, exponential mode of population growth with a finite ratio of the sexes. They also make a comparison of linear and nonlinear demographic models [128].

Hadeler, Waldstätter, and Wörz-Busekros in [43] later revisited Kendall's marriage model and carried out its systematic analysis. The homogeneity property of the marriage function is actually stressed in this work, and the model is analyzed in the general framework of homogeneous dynamical systems [41, 40, 42]. Their model differs from Kendall's in that it assumes different birth and death rates for females and males and involves a nonzero divorce rate for couples. The number of singles increases because of births (which may occur only in couples), death of a partner in a couple, and separation; it decreases because of natural death or marriage. Couples are only produced by marriage and destroyed by the death of one or both partners or by separation (including divorce). Thus, the dynamical system constituting the model is

$$\begin{cases} \dfrac{ds_f}{dt} = (\beta_f + \mu_m + \sigma)\, c - \mu_f s_f - \mathcal{M}(s_f, s_m), \\[2ex] \dfrac{ds_m}{dt} = (\beta_m + \mu_f + \sigma)\, c - \mu_m s_m - \mathcal{M}(s_f, s_m), \\[2ex] \dfrac{dc}{dt} = -(\mu_f + \mu_m + \sigma)\, c + \mathcal{M}(s_f, s_m). \end{cases}$$

Here s_f, s_m, and c stand, respectively, for the numbers of single females, single males, and couples (pairs); β_f and β_m represent the birth rates of females and males from couples; μ_f and μ_m represent the death rates of females and males; σ is the separation (including divorce) rate for couples; and $\mathcal{M}(s_f, s_m)$ is the marriage function.

In their analysis the authors observe that the homogeneity property of the nonlinearity leads to the fact that the system supports exponential solutions for a wide variety of choices of the parameters rather than time-independent steady states. They show existence of an exponential solution with a common growth rate for males and females and establish conditions for its global stability. The growth rate of the population is determined as the unique real solution λ of the nonlinear equation

$$\mathcal{M}\left(\frac{\beta_f}{\mu_f + \lambda} - 1, \frac{\beta_m}{\mu_f + \lambda} - 1\right) = \mu_f + \mu_m + \sigma + \lambda.$$

A discrete version of the model discussed by Hadeler et al. has been recently introduced by Castillo-Chávez et al. [17]. With the same notation as above, the model is governed by the difference equations

$$\begin{cases} s_f(t+1) = (\beta_f \mu_f \mu_m + (1 - \mu_m)\mu_f + (1 - \sigma))\, c + \mu_f s_f - \mathcal{M}(s_f, s_m), \\[2ex] s_m(t+1) = (\beta_m \mu_f \mu_m + (1 - \mu_f)\mu_m + (1 - \sigma))\, c - \mu_m s_m - \mathcal{M}(s_f, s_m), \\[2ex] c(t+1) = \mu_f \mu_m \sigma\, c + \mathcal{M}(s_f, s_m), \end{cases}$$

where μ_f and μ_m are, respectively, the proportions of females and males who survive to the next time period, and couples separate with probability $1 - \sigma$. The authors establish in [17] the existence of growing (or decaying) solutions, which in the discrete case have the form of geometric sequences. In contrast with the continuous case, the question of their stability is still open.

Actually, existence of exponential solutions (respectively, geometric) is a typical aspect of linear systems and, in fact, the nonlinearity due to the marriage function does not take into account the effects of crowding that result in logistic effects. In this direction, Castillo-Chávez and Huang [14] analyze a similar model, which does take into account crowding in the population. These effects are incorporated in the model via density-dependent birth and death rates. Here exponential solutions no longer occur, and the authors establish the existence of a unique steady state and its stability under suitable assumptions on the parameters.

The models discussed above—based on ordinary differential equations—have significantly advanced our understanding of population processes and the mathematics necessary to describe them. Although these models are relatively well understood and constitute a central element in our knowledge about pair formation, they fall short of building a two-sex theory and answering many demographic questions since they neglect an important demographic characteristic, namely the age structure.

Demographic literature abounds in age-structured two-sex models both continuous and discrete. Conventional life tables have been extended to two-sex populations in [109, 111, 113]. Among the continuous models there seem to be two basic approaches. The first one considers mostly integral equations (Lotka's approach) and aims at constructing appropriate birth functions [24, 26, 25, 27, 92, 93, 94]. The second one builds the two-sex model on the basis of the McKendrick–von Foerster one-sex model, which leads to the problem of formulating a realistic marriage function. The approaches are similar [26].

An age-structured extension of the first of Kendall's models was proposed by Nathan Keyfitz in 1972 [69]. This model is based on the two variables $u_f(x,t)$ and $u_m(y,t)$, which, respectively, are the age densities of the females and males at time t. Then, within the framework of the McKendrick–von Foerster model, we have the system

$$\begin{cases} \dfrac{\partial u_f}{\partial t} + \dfrac{\partial u_f}{\partial x} = -\mu_f(x,t)u_f(x,t), \\[2ex] \dfrac{\partial u_m}{\partial t} + \dfrac{\partial u_m}{\partial y} = -\mu_m(y,t)u_m(y,t), \\[2ex] u_f(0,t) = \dfrac{1}{M(t)+F(t)} \displaystyle\int_0^\infty \int_0^\infty \beta(x,y)u_f(x,t)u_m(y,t)\,dxdy, \\[2ex] u_m(0,t) = su_f(0,t), \end{cases} \tag{2.1}$$

where $F(t)$ and $M(t)$ are the total number of females and males, respectively,

$$F(t) = \int_0^\infty u_f(x,t)dx, \quad M(t) = \int_0^\infty u_m(y,t)dy,$$

and s is the sex ratio at birth assumed constant in time.

Keyfitz's is one of the simplest nonlinear two-sex age-structured models. Just as with the one-sex model, the main goal of its analysis is determining the growth rate through a characteristic equation, which in this case has the form

$$1 = \frac{s}{\hat{\pi}_f(\lambda) + s\hat{\pi}_m(\lambda)} \int_0^\infty \int_0^\infty \beta(x,y)e^{-\lambda(x+y)}\pi_f(x)\pi_m(y)\,dxdy. \tag{2.2}$$

Actually, this equation cannot be treated by the simple methods used in the one-sex cases since

$$\hat{\pi}_f(\lambda) = \int_0^\infty \pi_f(x)e^{-\lambda x}dx$$

and $\hat{\pi}_m(\lambda)$ depend on λ. Consequently, the expression on the right-hand side of (2.2) is not a monotone function of λ. One of the striking differences with the one-sex case is that this equation does not necessarily have a unique real solution. In fact it is not hard to create examples with multiple real solutions. For example, if we take

$$\beta(x, y) = \frac{1+s}{0.1s} xye^{-0.01x}e^{-0.01y}, \quad \mu_f(x, t) = \mu_m(y, t) = 0.1,$$

the characteristic equation (2.2) has two real and two complex solutions. The real solutions are $\lambda_1 = 2.041$ and $\lambda_2 = -0.0999999$. This simple example demonstrates that the presence of both sex and age structure in the models can potentially lead to much more complicated dynamics than either one without the other.

Discrete models have also been formulated extending the Leslie theory to the two-sex interaction. Perhaps the most interesting among the discrete models are the birth matrix-mating rule (BMMR) and the birth matrix-mating rule with persistent unions (BMMRPU) developed by Pollak [100, 101, 99, 102] (also see [19] for a general presentation of age-structured discrete population models). In that series of papers Pollak develops the framework of the BMMR. This model is fully discrete and considers a population structured by age, sex, and marital status and in its most general form with persistent unions (BMMRPU) is similar to a discrete analogue of the Fredrickson–Hoppensteadt model discussed in the next section. Pollak takes the mating rule to be a nonlinear function assumed to satisfy the usual properties of the mating function [87]. To introduce the model we denote by $F_i(t)$ and $M_i(t)$, respectively, the number of females and the number of males of age i at time t. Furthermore, the numbers of singles of age i at time t is denoted correspondingly by $F_i^s(t)$ and $M_i^s(t)$, and the number of unions by u_{ij}. μ_i^f and μ_i^m are the death probabilities for females and males of age i, σ_{ij} is the probability of separation of a couple, and β_{ij} is the birth rate per couple with spouses of ages i and j. Then, the model is given by the system

$$\begin{cases} F_i(t+1) = (1-\mu_{i-1}^f)F_{i-1}(t), \\ M_i(t+1) = (1-\mu_{i-1}^m)M_{i-1}(t), \\ u_{ij}(t+1) = (1-\sigma_{ij})(1-\mu_i^f)(1-\mu_j^m)u_{i-1,j-1}(t) \\ \qquad\qquad\qquad +\mathcal{M}(F_i^s(t+1), M_j^s(t+1)), \end{cases} \tag{2.3}$$

where the boundary conditions are given by

$$\begin{cases} F_1(t+1) = (1-\gamma)\sum_{i=1}^{n}\sum_{j=1}^{n} b_{ij}u_{ij}(t), \\ M_1(t+1) = \gamma\sum_{i=1}^{n}\sum_{j=1}^{n} b_{ij}u_{ij}(t), \\ u_{1,j}(t+1) = u_{i,1}(t+1) = 0, \end{cases} \tag{2.4}$$

with γ being the proportion of males among the newborn. The number of singles in each age–sex category is given by

$$\begin{cases} F_i^s(t+1) = F_i(t+1) - \sum_k (1-\sigma_{ij})(1-\mu_i^f)(1-\mu_j^m)u_{i-1,k-1}(t), \\ M_i^s(t+1) = M_i(t+1) - \sum_k (1-\sigma_{ij})(1-\mu_i^f)(1-\mu_j^m)u_{k-1,j-1}(t). \end{cases} \tag{2.5}$$

For the model with persistent unions Pollak gives the idea behind a proof of existence of an equilibrium age distribution, but he suggests that further work is needed to determine the stability properties of this solution.

2.2 The Fredrickson–Hoppensteadt model

A two-sex age-structured population model, following the framework of the second model by Kendall, was proposed by Fredrickson [32] and reintroduced by Hoppensteadt [48]. This model—which we discuss here—consists of McKendrick-type equations for the densities of females and males. The total birth rates for girls and boys, however, depend on the density of the couples. That is, the model assumes that births occur only in married couples. The model is described below and is based on the following state variables:

$u_f(x,t)$:	density of females of age x at time t;
$u_m(y,t)$:	density of males of age y at time t;
$c(x,y,t)$:	density of couples with wife of age x and husband of age y at time t.

The dynamics of the females and males are given by the equations

$$\begin{cases} \dfrac{\partial u_f}{\partial t} + \dfrac{\partial u_f}{\partial x} = -\mu_f(x,t)u_f(x,t), & x > 0, t > 0, \\ u_f(0,t) = B_f(t) = \displaystyle\int_0^\omega \int_0^\omega \beta_f(x,y,t)c(x,y,t)\,dxdy, & t > 0, \\ u_f(x,0) = u_f^0(x), & x \geq 0, \end{cases} \tag{2.6}$$

$$\begin{cases} \dfrac{\partial u_m}{\partial t} + \dfrac{\partial u_m}{\partial y} = -\mu_m(y,t)u_m(y,t), & y > 0, t > 0, \\ u_m(0,t) = B_m(t) = \displaystyle\int_0^\omega \int_0^\omega \beta_m(x,y,t)c(x,y,t)\,dxdy, & t > 0, \\ u_m(y,0) = u_m^0(y), & y \geq 0, \end{cases} \tag{2.7}$$

where

$\mu_f(x,t)$	is the age-specific death rate for females;
$\mu_m(y,t)$	is the age-specific death rate for males;
$\beta_f(x,y,t)$	is the age-specific birth rate for females;
$\beta_m(x,y,t)$	is the age-specific birth rate for males;
$u_f^0(x)$	is the initial age density of females;
$u_m^0(y)$	is the initial age density of males;
ω	is the maximal age.

The dynamics of the couples is given by a similar equation:

$$\begin{cases} \dfrac{\partial c}{\partial t} + \dfrac{\partial c}{\partial x} + \dfrac{\partial c}{\partial y} = -\sigma(x,y,t)c(x,y,t) \\ \qquad\qquad + \mathcal{M}(x,y,t;s_f,s_m), \quad x>0,\ y>0,\ t>0, \\ c(x,0,t) = c(0,y,t) = 0, \qquad x\geq 0,\ y\geq 0,\ t>0, \\ c(x,y,0) = c^0(x,y), \qquad x\geq 0,\ y\geq 0, \end{cases} \tag{2.8}$$

where

$c^0(x,y)$	is the initial density of couples;
$\sigma(x,y,t)$	is the separation rate for couples;
$s_f(x,t)$	is the density of single females;
$s_m(y,t)$	is the density of single males.

Here $\mathcal{M}(x,y,t;s_f,s_m)$ is the *marriage function* and it depends on the densities of single males and single females. Its value gives the density of couples formed at time t with wife of age x and husband of age y. The form and the properties of the marriage function are discussed in more detail in section 2.5.

We note that

$$\sigma(x,y,t) = \mu_f(x,t) + \mu_m(y,t) + \delta(x,y,t), \tag{2.9}$$

that is, the separation rate is due to death and divorce. The distributions of single females and males that enter in the marriage function can be expressed as the differences of the total age distributions for the corresponding sex and those of married individuals. The former are given by the primary variables $u_f(x,t)$ and $u_m(y,t)$, respectively. The distribution of married females is given by integrating the distribution of couples over all ages of the male partner,

$$\int_0^\omega c(x,y,t)\,dy,$$

while that of married males is given by integrating the distribution of couples over all ages of the female partner,

$$\int_0^\omega c(x,y,t)\,dx.$$

Therefore, we have the following explicit expressions for the distributions of single females and males:

$$\begin{cases} s_f(x,t) = u_f(x,t) - \displaystyle\int_0^\omega c(x,y,t)\,dy, \\ s_m(y,t) = u_m(x,t) - \displaystyle\int_0^\omega c(x,y,t)\,dx. \end{cases} \tag{2.10}$$

As far as demography is concerned, there are two central issues raised by this model. First, how does the marriage function look? This question has been a topic of discussion ever since the model was proposed. In fact $\mathcal{M}(x,y,t;s_f,s_m)$ is a central ingredient of this model since it describes the mode of interaction between females and males that represents the main step for reproduction. In section 2.5 we shall consider constitutive aspects regarding this function and we shall discuss the basic problems and achievements related to it. The second question is whether this model (taken with time-independent birth and death rates) indeed extends the one-sex stable population theory and in what sense. This question is partially discussed in Chapter 5. It is our belief that the results presented there may be viewed as a starting point of a two-sex stable population theory. When this model was published in 1975, it was considered inaccessible for mathematical analysis. Currently it is receiving considerable attention due to its central position (it models stable contact structures) in investigating sexually transmitted diseases and, in particular, AIDS. The Fredrickson–Hoppensteadt model is our main concern in this book.

2.3 Hadeler's model

An extension of the previous model was proposed by Hadeler in [38], considering a further structure in the couples. Namely, a new independent variable z is introduced to represent the duration of the marriage (the age of the couple) so that couples are described by a density function $c(x,y,z,t)$ of three age variables: x (the age of the female), y (the age of the male), and z (the age of the couple). The function $c(x,y,z,t)$ can also be viewed as the density at time t of couples with female of age x and male of age y who married at time $t-z$. In this model the state of the population is described through the densities $s_f(x,t)$ (single females) and $s_m(y,t)$ (single males) rather than the total number of individuals of each sex. The equations of the model are

$$\begin{aligned}
&\frac{\partial s_f}{\partial t} + \frac{\partial s_f}{\partial x} + \mu_f(x)\, s_f + \int_0^\infty \mathcal{M}(x, y, t; s_f, s_m)\, dy \\
&\qquad - \int_0^\infty \int_0^\infty [\mu_m(y) + \delta(x, y, z)]\, c(x, y, z, t)\, dy\, dz = 0, \\
&\frac{\partial s_m}{\partial t} + \frac{\partial s_m}{\partial y} + \mu_m(y) s_m + \int_0^\infty \mathcal{M}(x, y, t; s_f, s_m)\, dx \\
&\qquad - \int_0^\infty \int_0^\infty \big[\mu_f(x) + \delta(x, y, z)\big]\, c(x, y, z, t)\, dx\, dz = 0, \\
&\frac{\partial c}{\partial t} + \frac{\partial c}{\partial x} + \frac{\partial c}{\partial y} + \frac{\partial c}{\partial z} + \big[\mu_f(x) + \mu_m(y) + \delta(x, y, z)\big]\, c = 0,
\end{aligned} \tag{2.11}$$

endowed with the boundary conditions

$$\begin{aligned}
s_f(0, t) &= \int_0^\infty \int_0^\infty \beta_f(x, y, z)\, c(x, y, z, t)\, dx\, dy\, dz, \\
s_m(0, t) &= \int_0^\infty \int_0^\infty \beta_m(x, y, z)\, c(x, y, z, t)\, dx\, dy\, dz,
\end{aligned}$$

and

$$\begin{aligned}
&c(0, y, z, t) = c(x, 0, z, t) = 0, \\
&c(x, y, 0, t) = \mathcal{M}(x, y, t; s_f, s_m), \\
&c(x, y, z, t) = 0 \quad \text{for} \quad z \geq \min\{x, y\}.
\end{aligned}$$

Here $\beta_f(x, y, z)$, $\beta_m(x, y, z)$, $\mu_f(x)$, $\mu_m(y)$, and $\delta(x, y, z)$ have the same meaning as in the Fredrickson–Hoppensteadt model, but β_f, β_m, and δ also depend on the variable z. Moreover, the boundary condition at $z = 0$ represents the fact that new couples are only produced by marriage. The last condition must be enforced since nobody can be married at or before birth.

We note that this model presents a fairly natural description of a sex-structured population but, adopting the variables s_f and s_m, the differential equations need to include complicated integral terms to account for the fact that the age densities of the individual sexes do not include individuals who have formed couples. Thus, system (2.11) is somewhat more complicated than the Fredrickson–Hoppensteadt system. Actually, the use of the total densities u_f and u_m produces a different—but equivalent—approach that consists of considering the age densities of the individual sexes as including individuals who have formed couples as well as those who have not—thus removing from the partial differential equations all the integral terms. This choice would simplify (2.11) into the system

$$\begin{cases} \dfrac{\partial u_f}{\partial t} + \dfrac{\partial u_f}{\partial x} + \mu_f(x)\,u_f = 0, \\[2ex] \dfrac{\partial u_m}{\partial t} + \dfrac{\partial u_m}{\partial y} + \mu_m(y)u_m = 0, \\[2ex] \dfrac{\partial c}{\partial t} + \dfrac{\partial c}{\partial x} + \dfrac{\partial c}{\partial y} + \dfrac{\partial c}{\partial z} + \big[\mu_f(x) + \mu_m(y) + \delta(x,y,z)\big]\,c = 0. \end{cases}$$

We finally note that, when the fertility and divorce rates—β_f, β_m, and δ—are independent of the age of the couple, z, then upon integration in z the model reduces to the Fredrickson–Hoppensteadt one with

$$c(x,y,t) = \int_0^\infty c(x,y,z,t)dz.$$

2.4 A model with births within and outside marriage

The Fredrickson–Hoppensteadt model is based on the explicit assumption that the population is actually monogamous [32]. Based on the couples dynamics, a couple, though not necessarily legally formed, is, however, the exclusive status that allows reproduction. That is, the model does not allow births outside marriage (or outside a couple).

The problem is made even more complicated if we want to use real-life data in the model. As a matter of fact, in order to have a detailed description of the joint distribution of births by the age of both parents, it is necessary to know the birth rates for girls and boys, β_f and β_m, for all births in the population combined. In real life, these data are usually impossible to obtain, since many birth certificates report the age of the mother but not that of the father. In many countries, however, including the United States, data concerning fertility are gathered separately for births from married and unmarried mothers. For married ones β_f and β_m are computed excluding births from unmarried mothers. Thus, a refinement of the model is possible by introducing two more birth rates—those for female and male offspring of unmarried mothers broken down by the age of the mother, respectively, $\beta_f^{(s)}$ and $\beta_m^{(s)}$. Then, we can easily express the total number of newborn females as the sum of those born from married mothers and those born from unmarried ones:

$$B_f(t) = \int_0^\omega \int_0^\omega \beta_f(x,y,t)\,c(x,y,t)\,dx\,dy + \int_0^\omega \beta_f^{(s)}(x)\,s_f(x,t)\,dx.$$

Similarly, we can express the total number of newborn males as the sum of those born from married mothers and those born from unmarried ones:

$$B_m(t) = \int_0^\omega \int_0^\omega \beta_m(x,y,t)\,c(x,y,t)\,dx\,dy + \int_0^\omega \beta_m^{(s)}(x)\,s_f(x,t)\,dx.$$

These constitutive forms of the birth rates would replace the corresponding expressions in (2.6) and (2.7) to obtain a modified model.

A major problem exists, however, with this model in that the two fertility rates for unmarried females ignore entirely the availability of males to impregnate them: the average number of offspring of each sex that an unmarried female will have is assumed to be the same in complete absence of males or in great abundance of them. Yet we observe that the number of matings and the productivity of offspring depend not only on the availability of females but also on that of males. As a consequence, this model can only work well for populations that are more or less persistent in the sense of stable population theory.

One way to account for the dependence of these fertility rates for unmarried females, $\beta_f^{(s)}$ and $\beta_m^{(s)}$, on the availability of males is to make them dependent on the size of the male population. Let

$$M(t) = \int_0^\omega u_m(y,t)\,dy$$

denote the size at time t of the total male population. We may then assume

$$\beta_f^{(s)} = \beta_f^{(s)}(x, M(t)) = \phi(M(t))\,\tilde{\beta}_f^{(s)}(x),$$

$$\beta_m^{(s)} = \beta_m^{(s)}(x, M(t)) = \phi(M(t))\,\tilde{\beta}_m^{(s)}(x),$$

where $\phi(M)$ is an increasing function with values in the interval $[0, 1]$ that describes the increase or reduction in the intrinsic fertility rates for unmarried females, $\tilde{\beta}_f^{(s)}$ and $\tilde{\beta}_m^{(s)}$, due to the availability of males for impregnating them, or lack thereof. Examples of such functions ϕ are reverse negative exponentials and Michaelis–Menten type functions

$$\phi(M) = 1 - e^{-\varepsilon M} \quad \text{or} \quad \phi(M) = \frac{M}{K + M}$$

for some positive constants ε, K.

2.5 The marriage function: Definition, properties, and examples

The significance of the marriage function emerged from the two-sex problem that we introduced in the previous sections. The question of its existence and its form began being widely discussed sometime in the early 1970s. It was interesting both from the point of view of the two-sex model and because it was expected to make sense of the male and female marriage rates [69]. Moreover, people were interested in forecasting marriages for different (mainly business-related) purposes. A general perception in demography is that the marriage system represents a market and is ruled by laws of competition. On a personal level marriage is an act expected to bring more comfort in life [117] and, on the contrary, health [75] and/or premarital childbearing [4] are some of the factors that can decrease one's chance for getting married (see also [3]).

Demographers have approached the problem by formulating sets of axioms that capture the basic requirements to be fulfilled by a marriage function. We adopt here a definition of the marriage function that follows the one given by McFarland [87]:

> *A* **marriage function** *is a function for predicting the number of marriages that will occur during a unit of time between males in particular age categories and females in particular age categories from knowledge of the numbers available in the various categories.*

Marriage is a complex socioeconomic process influenced by many factors. Some of them are dominating perceptions and rituals, religion, health, economy, educational status, racial and ethnic interactions, and age and sex composition. As can be seen from the definition above, in our context it is assumed that age and sex composition are the only essential properties that should be explicitly taken into account in modeling marriages.

In this section and in the rest of this chapter we present and discuss some of the major marriage models known today and accumulated after Kendall during the years of intense discussion of the problem. The purpose is not so much to acquaint the reader with these models, but rather to create a feeling about the diversity of methods used in their derivation, as well as the differences in the results obtained and the lack of tools for their comparison.

We begin by listing the formal requirements mentioned above, which are widely accepted by demographers and mathematicians. Actually, these attributes of the marriage function, which will be referred to as *properties of the marriage function*, are motivated by the axioms summarized by McFarland [87] and J. H. Pollard [106] as a result of discussions and on the basis of the predominant opinions. The formulation of the axioms in [87, 106] uses the discrete-age framework, while our formulation concerns the marriage function $\mathcal{M}(x, y, t; s_f, s_m)$ introduced in the continuous model by Fredrickson and Hoppensteadt presented in section 2.2. The properties of the marriage function in a continuous setting have also been discussed by Inaba [59].

(P1) The rate of marriage is nonnegative and is defined on nonnegative arguments:

$$\mathcal{M}(x, y, t; s_f, s_m) \geq 0$$

for $\quad x \geq 0,\ y \geq 0,\ t \geq 0,\ s_f(x, t) \geq 0,\ s_m(y, t) \geq 0.$

(P2) The marriage function vanishes if the singles of at least one of the two sexes are absent:

$$\mathcal{M}(x, y, t; 0, s_m) = \mathcal{M}(x, y, t; s_f, 0) = 0.$$

This property reflects the assumption that marriage occurs only between individuals of different sexes.

(P3) The marriage function is homogeneous with respect to s_f and s_m:

$$\mathcal{M}(x, y, t; \alpha s_f, \alpha s_m) = \alpha \mathcal{M}(x, y, t; s_f, s_m) \quad \text{for} \quad \alpha \geq 0.$$

This is the so-called homogeneity property and reflects the concept that, if the total population of singles increases α times and the sex ratio is preserved, then the births should also increase α times. Since the births and marriages are linearly related in this model, the same is true for the marriage function.

(P4) The number of marriages that involve individuals of a certain age and sex should be smaller than the total number of single individuals of that same age and sex:

$$\int_0^{\omega} \mathcal{M}(x, y, t; s_f, s_m)\, dy \leq s_f(x, t),$$

$$\int_0^{\omega} \mathcal{M}(x, y, t; s_f, s_m)\, dx \leq s_m(y, t).$$

This is actually a natural consistency condition.

(P5) An increase of the availability of singles occurring exclusively within a given age interval produces an increase of the total number of marriages occurring in that same age interval:

for any pair of intervals $[x_1, x_2]$, $[y_1, y_2]$ we have

$$\int_{y_1}^{y_2} \int_{x_1}^{x_2} \mathcal{M}(x, y, t; s_f, s_m)\, dx dy \leq \int_{y_1}^{y_2} \int_{x_1}^{x_2} \mathcal{M}(x, y, t; s'_f, s'_m)\, dx dy$$

if

$$s_f(x, t) \leq s'_f(x, t) \text{ for } x \in [x_1, x_2] \text{ and } s_f(x, t) = s'_f(x, t) \text{ for } x \notin [x_1, x_2],$$

$$s_m(y, t) \leq s'_m(y, t) \text{ for } y \in [y_1, y_2] \text{ and } s_m(y, t) = s'_m(y, t) \text{ for } y \notin [y_1, y_2].$$

In particular, this property implies that

$$\int_0^{\omega} \int_0^{\omega} \mathcal{M}(x, y, t; s_f, s_m)\, dx dy \leq \int_0^{\omega} \int_0^{\omega} \mathcal{M}(x, y, t; s'_m, s'_f)\, dx dy$$

for $s_f(x, t) \leq s'_f(x, t)$ and $s_m(y, t) \leq s'_m(y, t)$.

That is, the *total* number of marriages increases when the number of single individuals increases.

(P6) The "marriage market" is competitive:

for any pair of intervals $[x_1, x_2]$, $[y_1, y_2]$ we have

$$\int_{y_1}^{y_2} \int_{x_1}^{x_2} \mathcal{M}(x, y, t; s_f, s_m)\, dx dy \geq \int_{y_1}^{y_2} \int_{x_1}^{x_2} \mathcal{M}(x, y, t; s'_f, s'_m)\, dx dy \tag{2.12}$$

if

$$s_f(x, t) = s'_f(x, t) \text{ for } x \in [x_1, x_2] \text{ and } s_f(x, t) \leq s'_f(x, t) \text{ for } x \notin [x_1, x_2],$$

$$s_m(y, t) = s'_m(y, t) \text{ for } y \in [y_1, y_2] \text{ and } s_m(y, t) \leq s'_m(y, t) \text{ for } y \notin [y_1, y_2].$$

In fact this condition imposes that, for each sex, the number of marriages within a given age interval is nonincreasing if the availability of singles increases in another age interval. Actually, the property above is expressed in a weak form because the inequality is not strict and a null effect is allowed. A stronger form of this property is introduced in [87] and [106]:

for some pair of intervals $[x_1, x_2]$, $[y_1, y_2]$ there is a positive constant C such that inequality (2.12) is strict for

$$s'_f(x,t) > s_f(x,t) > C \text{ and/or } s'_m(y,t) > s_m(y,t) > C.$$

We now list five functions that were first proposed (together with their age-independent counterparts) and investigated as possible candidates for a marriage function:

$$\mathcal{M}(x,y,t;s_f,s_m) = \rho(x,y)s_f(x,t): \qquad \text{female dominance;} \tag{E1}$$

$$\mathcal{M}(x,y,t;s_f,s_m) = \rho(x,y)s_m(y,t): \qquad \text{male dominance;} \tag{E2}$$

$$\mathcal{M}(x,y,t;s_f,s_m) = 2\rho(x,y)\frac{s_f(x,t)s_m(y,t)}{s_f(x,t)+s_m(y,t)}: \qquad \text{harmonic mean;} \tag{E3}$$

$$\mathcal{M}(x,y,t;s_f,s_m) = \rho(x,y)\sqrt{s_f(x,t)s_m(y,t)}: \qquad \text{geometric mean;} \tag{E4}$$

$$\mathcal{M}(x,y,t;s_f,s_m) = \rho(x,y)\min(s_f(x,t),s_m(y,t)): \quad \text{minimum function.} \tag{E5}$$

These marriage models, in the form above and in their discrete form, have been the most discussed examples and the starting point of further generalizations. In his paper [38] Hadeler introduces the following family of marriage functions in the form of generalized weighted means of single female and single male densities:

$$\mathcal{M}(x,y,t;s_f,s_m) = \rho(x,y)\left[\beta s_f(x,t)^\alpha + (1-\beta)s_m(y,t)^\alpha\right]^{1/\alpha}. \tag{2.13}$$

The family above includes all examples (E1)–(E5) as particular cases. In fact the cases $\beta = 1$ and $\beta = 0$ correspond, respectively, to the female and male dominance functions (E1) and (E2). The case $0 < \beta < 1, \alpha \to -\infty$, corresponds to the minimum function (E5), while $\alpha = -1$ corresponds to the sex-biased harmonic mean function

$$\mathcal{M}(x,y,t;s_f,s_m) = \rho(x,y)\frac{s_f(x,t)s_m(y,t)}{\beta s_f(x,t)+(1-\beta)s_m(y,t)}, \tag{E6}$$

which is the true harmonic mean when $\beta = \frac{1}{2}$. Finally, $\alpha \to 0^-$ corresponds to the sex-biased geometric mean function

$$\mathcal{M}(x,y,t;s_f,s_m) = \rho(x,y)(s_f(x,t))^\beta(s_m(y,t))^{1-\beta}, \tag{E7}$$

which is the true geometric mean when $\beta = \frac{1}{2}$.

A further extension of (2.13) was introduced in [84], taking into account preferences among individuals:

$$\mathcal{M}(x,y,t;s_f,s_m) = \left[\beta g(x,y)^\alpha s_f(x,t)^\alpha + (1-\beta)h(x,y)^\alpha s_m(y,t)^\alpha\right]^{1/\alpha}. \tag{2.14}$$

In this form of $\mathcal{M}(x, y, t; s_f, s_m)$ the *preference distributions* $g(x, y)$ and $h(x, y)$ are used with the following meaning. If we know that a female of age x considers males of age y as possible partners with probability $g(x, y)$, then we call g the preference distribution of females. Similarly, if we know that a male of age y considers females of age x as possible partners with probability $h(x, y)$, then we call h the preference distribution of males. When g and/or h are constant, there is no preference; when they are equal, $\mathcal{M}$ is given by (2.13) with $\rho = g = h$.

The basic examples above have withstood a lot of criticism, but they and their extensions are still at the center of discussion for marriage models. Parlett [98] criticizes the lack of additivity properties, observing that the number of marriages in a five-year age group should be the sum of the corresponding number of marriages in one-year age groups. To further discriminate among the candidates McFarland tested them against the formal properties. He established that none of the five functions satisfies all six conditions (see Table 2.1 where property (P6) is intended in the strong form). However, the harmonic mean satisfies all of the conditions imposed on the marriage function except the last one (although Schoen [112] argues that the lack of sensitivity of the harmonic mean to the number of single females or males in the other age groups is only apparent). To remedy that situation the following generalization of (E3) was proposed [39]:

$$\mathcal{M}(x, y, t; s_f, s_m) = 2\rho(x, y)\frac{s_m(x, t)s_f(y, t)}{\int_0^\omega s_m(x, t)\,dx + \int_0^\omega s_f(y, t)\,dy}. \tag{E8}$$

Table 2.1. *Properties satisfied by various marriage functions.*

	E1	E2	E3	E4	E5	E6	E7	E8
P1	YES	YES	YES	YES	YES	YES	YES	YES
P2	NO	NO	YES	YES	YES	YES	YES	YES
P3	YES	YES	YES	YES	YES	YES	YES	YES
P4	NO	NO	YES	NO	YES	YES	YES	YES
P5	YES	YES	YES	YES	YES	YES	YES	YES
P6	NO	NO	NO	NO	NO	NO	NO	YES

Some authors also argue (see [105, 60, 17]) that the marriage function should depend not only on the single males and females but also on the couples. An extension of this form was considered by Inaba in [60]:

$$\mathcal{M}(x, y, t; s_f, s_m, c) = 2\rho(x, y)\frac{s_m(x, t)s_f(y, t)}{\int_0^\omega s_m(x, t)\,dx + \int_0^\omega s_f(y, t)\,dy + \int_0^\omega \int_0^\omega c(x, y, t)\,dxdy}. \tag{E9}$$

This last form is intended to take into account the fact that the contacts among individuals occur not only among singles but also in the entire population.

As we can see from the variety of marriage functions that have been proposed and the discussion about them (later we will consider more examples in correspondence with some specific modeling aspects), properties (P1)–(P6) are rather general and the mathematical expressions of such a function may be of many types. In fact these properties leave the door open to general functional dependence of $\mathcal{M}(x, y, t; s_f, s_m)$ upon the age distributions of singles. Moreover, we have seen that the marriage function may also depend on other variables, such as the couples distribution, and it may depend as well on the total population (both sexes included).

To base the mathematical treatment on a systematic mathematical formulation we shall restrict the choice of the marriage function $\mathcal{M}(x, y, t; s_f, s_m)$ to special but significant forms that we divide into three classes and formalize in the following way.

Marriage functions of the first kind:

The marriage functions of this class have the form

$$\mathcal{M}(x, y, t; s_f, s_m) = \mathcal{F}\left[x, y, t, s_f(x, t), s_m(y, t)\right],$$

where $\mathcal{F}[x, y, t, f, m]$ is a function of the five real variables (x, y, t, f, m) and is defined in the hypercube $[0, \omega]^2 \times [0, \infty]^3$. More assumptions on this function must be made to match the properties (P1)–(P5) discussed above. All the examples (E1)–(E7) belong to this class.

Marriage functions of the second kind:

In this case we consider the function $\mathcal{F}[x, y, t, f, m, \vec{f}, \vec{m}]$, where we have used the vector notation $\vec{f} \equiv (f_1, \ldots, f_N)$ and $\vec{m} \equiv (m_1, \ldots, m_N)$, so that the function $\mathcal{F}$ is defined on $[0, \omega]^2 \times [0, \infty]^{3+2N}$:

$$\mathcal{M}(x, y, t; s_f, s_m) = \mathcal{F}\left[x, y, t, s_f(x, t), s_m(y, t), \vec{\Gamma}_f(y, t), \vec{\Gamma}_m(x, t)\right],$$

where $\vec{\Gamma}_f(y, t) \equiv (\Gamma_f^1(y, t), \ldots, \Gamma_f^N(y, t))$, $\vec{\Gamma}_m(x, t) \equiv (\Gamma_m^1(x, t), \ldots, \Gamma_m^N(x, t))$, and

$$\Gamma_f^i(y, t) = \int_0^\omega \gamma_f^i(y, \sigma) s_f(\sigma, t) d\sigma, \quad \Gamma_m^i(x, t) = \int_0^\omega \gamma_m^i(x, \sigma) s_m(\sigma, t) d\sigma$$

for $i = 1, \ldots, N$, with $\gamma_f^i(y, \sigma)$ and $\gamma_m^i(x, \sigma)$ being given weight functions.

This class is considered in order to include integral terms with some different particular weights that allow us to select significant ages of the densities. In particular, they may model the preference of one age class for a different age class in the other sex. Special cases include weights of the proportionate form

$$\gamma_f^i(x, \sigma) = h_f^i(x) k_f^i(\sigma), \quad \gamma_m^i(y, \sigma) = h_m^i(y) k_m^i(\sigma),$$

as well as of the convolution-type form

$$\gamma_f^i(x, \sigma) = g_f^i(x - \sigma), \quad \gamma_m^i(x, \sigma) = g_m^i(x - \sigma).$$

The weights of convolution type are particularly apt to capture the dependence of preferences on the age difference. The example (E8) belongs to this class and it may be extended to the form

$$\mathcal{M}(x, y, t; s_f, s_m) \tag{E10}$$
$$= \rho(x, y) \frac{s_m(x, t) s_f(y, t)}{\left[\beta \left(\int_0^\omega \gamma_f(x, \sigma) s_m(\sigma, t)\, d\sigma\right)^\alpha + (1 - \beta) \left(\int_0^\omega \gamma_m(y, \sigma) s_f(\sigma, t)\, d\sigma\right)^\alpha\right]^{1/\alpha}},$$

whose discrete form, with $\alpha = 1$ and $\beta = \frac{1}{2}$, was considered by J. H. Pollard in [106]. Of course the functions of the first kind are a subclass of the present one.

Marriage functions of the third kind:

This class further extends the previous two classes by allowing dependence of the marriage function on the density of couples. Here we consider the function $\mathcal{F}[x, y, t, f, m, c, \vec{f}, \vec{m}, \vec{c}, \vec{c}_f, \vec{c}_m]$, defined on $[0, \omega]^2 \times [0, \infty]^{4+5N}$, together with the weight functions $\gamma_f^i(y, \sigma), \gamma_m^i(x, \sigma), \gamma_c^i(x, y), \gamma_{c,f}^i(y), \gamma_{c,m}^i(x)$ $(i = 1, \dots, N)$, and we define

$$\mathcal{M}(x, y, t; s_f, s_m, c)$$

$$= \mathcal{F}\Big[x, y, t, s_f(x,t), s_m(y,t), c(x,y,t), \vec{\Gamma}_f(y,t), \vec{\Gamma}_m(x,t), \vec{\Gamma}_c(t), \vec{\Gamma}_{c,f}(x,t), \vec{\Gamma}_{c,m}(y,t)\Big],$$

where, in addition to the previous notation, we have set

$$\vec{\Gamma}_c(t) \equiv \left(\Gamma_c^1(t), \dots, \Gamma_c^N(t)\right), \quad \Gamma_c^i(t) = \int_0^\omega \int_0^\omega \gamma_c^i(x, y) c(x, y, t)\, dx\, dy,$$

$$\vec{\Gamma}_{c,f}(x,t) \equiv \left(\Gamma_{c,f}^1(x,t), \dots, \Gamma_{c,f}^N(x,t)\right), \quad \Gamma_{c,f}^i(x,t) = \int_0^\omega \gamma_{c,f}^i(y) c(x, y, t)\, dy,$$

$$\vec{\Gamma}_{c,m}(y,t) \equiv \left(\Gamma_{c,m}^1(y,t), \dots, \Gamma_{c,m}^N(y,t)\right), \quad \Gamma_{c,m}^i(y,t) = \int_0^\omega \gamma_{c,m}^i(x) c(x, y, t)\, dx.$$

With the three classes defined above we have not exhausted all possible interesting examples, but at least we have incorporated a few cases that may be the basis for further modeling. Concerning the functions $\mathcal{F}$ occurring in the definition of the above classes, we will make suitable assumptions when needed. However, we note that basic assumptions will reflect the nature of examples (E1)–(E10) presented above.

Because it fitted the desired properties best, the harmonic mean (and its generalization (E8)) was the favorite among the first five. However, when the five functions are tested against real data [69, 110], the error between the projections of either one of them and the actual marriage data is much bigger than the differences in their projections (and that is always the case regardless of the method or the norm used for the comparison). So it seems that comparison on the basis of real data cannot single out one function as the best performer. Moreover, computationally, the harmonic mean seems not to perform according to expectations (this observation was first made by Keyfitz [69]) and, in fact, is often among the worst of the five. The fact that these functions are too simple to describe as complicated a process as marriage is believed to be one possible reason.

A few numerical methods were created to project the marriage matrix (that is, a matrix whose (i, j)th entry is the number of marriages between a bride in the ith age group and a groom in the jth). The two leading ones are McFarland's iterative adjustment model [88] and Henry's panmictic circles model [47]. McFarland's method is based on a procedure designed to change the row and column totals of a matrix while preserving all of its cross-product ratios. Henry's solution is based on the assumption that people meet and get married within one or more overlapping "circles" that are homogeneous with respect to all population characteristics and in which marriage occurs randomly. The two methods numerically manipulate the entries of the current marriage matrix and the arrays of single males and females to obtain a new (projected) marriage matrix. They indeed improve slightly the projection of marriages but are also difficult to compare with each other. We believe that their improved performance in comparison with the functional models is not due so much to their more complicated nature but rather to their more complete use of the marriage data.

Caswell [18] brings out another possible reason why data cannot effectively distinguish among the various marriage models. He points out that the logical differences among the functions are made on the basis of extremal situations. For example, the male dominance function is clearly not valid when there are no females in the society, or the geometric mean will project large numbers of marriages if one of the sexes is in abundance compared to the other, and so on. However, in reality the sex ratio in the human population is quite close to one and the real data reflect that fact.

2.6 The marriage function as a part of the problem of modeling human sexual interactions

In the previous section we introduced and discussed the main properties of the marriage function and presented some examples that fit to a certain extent into the framework defined by these properties. In this section we discuss some constitutive arguments that produce these examples from the point of view of the mechanisms that stand behind them. In fact we describe how individuals of different sexes interact and then produce specific forms of the marriage function as a result of a mixing/mating process.

Interest in this process and, consequently, in marriage models was renewed in the 1980s as the work on modeling scenarios of AIDS progression began occupying a central place. In fact it became apparent that on the one hand knowledge of the contact dynamics of married people is important since this social group is subjected to a significantly lower risk of contracting AIDS, and on the other hand the problem itself is a special case of the problem of modeling human sexual interaction [7, 9, 11, 39, 29].

The main progress here is due to Carlos Castillo-Chávez and coworkers, who, in a series of papers ([12]–[11] and [15]–[17]), develop a formalism in both discrete and structured cases and apply it to various areas—anthropology, demography, epidemiology, food web dynamics, and others. As a byproduct of this axiomatic framework, one can obtain an age-structured marriage function [53, 13]. We will follow their formulation to discuss the constitutive definition of marriage functions in the context of an age-structured two-sex population.

We begin by introducing the basic variables we use to describe the mechanism. Let

$$C_f(x,t) = \text{ the per capita marriage rate of females of age } x \text{ at time } t,$$

$$C_m(y,t) = \text{ the per capita marriage rate of males of age } y \text{ at time } t.$$

Thus, $C_f(x,t)s_f(x,t)dx$ is the number of marriages of single females of age x per unit time (respectively, $C_m(y,t)s_m(y,t)dy$ is the number of marriages of single males of age y per unit time). Then, let

$p(x, y, t) =$ the probability that a female of age x marries a male of age y given that she marries,

$q(x, y, t) =$ the probability that a male of age y marries a female of age x given that he marries.

As defined above, the parameters of the model should satisfy a natural set of properties at all times:

$$p(x, y, t) \geq 0 \quad \text{and} \quad q(y, x, t) \geq 0, \tag{A1}$$

$$\int_0^\infty p(x, y, t)\, dy = 1 \text{ for all } x, \qquad \int_0^\infty q(y, x, t)\, dx = 1 \text{ for all } y, \tag{A2}$$

$$p(x, y, t)C_f(x, t)s_f(x, t) = q(y, x, t)C_m(y, t)s_m(y, t) \text{ for all } x \text{ and } y. \tag{A3}$$

The third axiom is the *axiom of conservation of number of interactions*, which roughly means that the number of females of age x who marry a male of age y should be equal to the number of males of age y who marry a female of age x.

Obviously, the number of women who marry at time t should be equal to the number of men who marry at time t so that the above parameters of the model in the current context must satisfy

$$\int_0^\infty C_f(x, t)s_f(x, t)\, dx = \int_0^\infty C_m(y, t)s_m(y, t)\, dy. \tag{2.15}$$

Actually, (2.15) can be formally obtained from (A2) and (A3) by integrating both sides of (A3) with respect to x and y.

Finally, in the above framework the marriage function can be expressed by one of the two forms

$$\begin{aligned}\mathcal{M}(x, y, t; s_f, s_m) &= p(x, y, t)C_f(x, t)s_f(x, t) \\ &= q(x, y, t)C_m(y, t)s_m(y, t).\end{aligned} \tag{2.16}$$

It is still necessary to specify the functions $C_f(x,t)$, $C_m(y,t)$, $p(x,y,t)$, and $q(x,y,t)$. A simplification occurs if we assume that the preference in the choice of a partner does not depend on one's own age. This amounts to assuming that the probabilities $p(x, y, t)$ and $q(x, y, t)$ are actually independent of x and y, respectively. In fact in this case we must have (see (A3) and (2.15))

$$p(x, y, t) = \bar{p}(y, t) = \frac{C_m(y, t)s_m(y, t)}{\int_0^\omega C_m(y, t)s_m(y, t)dy} \tag{2.17}$$

and

$$q(x, y, t) = \bar{q}(x, t) = \frac{C_f(x, t)s_f(x, t)}{\int_0^\omega C_f(x, t)s_f(x, t)dx}. \tag{2.18}$$

That is, $\bar{p}(y,t)$ (respectively, $\bar{q}(x,t)$) is the ratio of marriages of males (respectively, females) of age y (respectively, x) to the total number of marriages.

To see how the mixing framework leads to marriage functions from the previous section, we give a form to $C_f(x,t)$ and $C_m(y,t)$, keeping in mind that they should satisfy (2.15). We consider the expressions

$$M_f(x) = \text{marriage inclination for a single female of age } x,$$

$$M_m(y) = \text{marriage inclination for a single male of age } y.$$

This parameter obviously reflects personal inclination, but also social habits and physiological aspects (sexual maturation). Consequently, the functions $M_f(x)$ and $M_m(y)$ are independent of the availability of single males or females. Let

$\mathcal{K}(N(t))$ = contact rate, i.e., the average number of contacts that a single individual realizes within a "socially active" population of size $N(t)$.

This rate may actually be the same for females and males. It is an increasing function of $N(t)$. Moreover,

$$\mathcal{K}(0) = 0 \quad \text{and} \quad \mathcal{K}(+\infty) = \mathcal{K}_\infty < \infty.$$

Here $N(t)$ may be assumed to be the total population,

$$N(t) = \int_0^\omega s_f(x,t)dx + \int_0^\omega s_m(y,t)dy + 2\int_0^\omega \int_0^\omega c(x,y,t)dxdy, \tag{2.19}$$

but also a weighted section of it,

$$N(t) = \int_0^\omega \gamma_f(x)s_f(x,t)dx + \int_0^\omega \gamma_m(y)s_m(y,t)dy + \int_0^\omega \int_0^\omega \gamma_c(x,y)c(x,y,t)dxdy. \tag{2.20}$$

Finally, we consider

$\dfrac{\int_0^\omega M_m(y)s_m(y,t)dy}{N(t)}$ = the probability that a given contact within the "socially active" population occurs with a single male inclined to marry.

Now we assume that the per capita marriage rate for a female of age x is the product of her inclination to marry and the probability that she has a contact with an appropriate partner. This leads to the following constitutive form for $C_f(x,t)$:

$$C_f(x,t) = M_f(x)\frac{\mathcal{K}(N(t))}{N(t)}\int_0^\omega M_m(y)s_m(y,t)dy.$$

Similarly, for $C_m(y,t)$ we obtain

$$C_m(y,t) = M_m(y)\frac{\mathcal{K}(N(t))}{N(t)}\int_0^\omega M_f(x)s_f(x,t)dx.$$

It is easy to see that with this constitutive form condition (2.15) is satisfied. Moreover, under (2.17) and (2.18), the marriage function has the form

$$\mathcal{M}(x, y, t; s_f, s_m) = M_f(x)M_m(y)\frac{\mathcal{K}(N(t))}{N(t)}s_f(x, t)s_m(y, t).$$

This is a marriage function of the third type and it satisfies properties (P1)–(P6). If the contact rate $\mathcal{K}$ is assumed to be constant (as is often assumed for large populations), then we have examples (E8)–(E10) depending on the particular choice of the weights in (2.20).

The separable case (2.17)–(2.18) models an age-structured, proportionately mixing, two-sex population. Castillo-Chávez and Busenberg [12] prove that all solutions (p, q) of (A1)–(A3) are multiplicative perturbations of this solution (called the Ross solution). The general form of the two-sex mixing functions depends on parameter functions that can be practically interpreted as *preferences.* In particular, if $g(x, y)$ is the preference (affinity) of females age x to males age y, $h(y, x)$ is the preference (affinity) of males of age y to females of age x, and

$$\begin{aligned}
R_f(x) &= 1 - \int_0^\infty \bar{p}(y)g(x, y)\,dy, \\
R_m(y) &= 1 - \int_0^\infty \bar{q}(x)h(y, x)\,dx, \\
V_f &= \int_0^\infty \bar{q}(x)R_f(x)\,dx, \\
V_m &= \int_0^\infty \bar{p}(y)R_m(y)\,dy
\end{aligned} \tag{2.21}$$

(see [53] for the physical interpretation of these quantities), then we can formulate the representation theorem for the two-sex mixing functions.

Theorem 2.1. *For each pair (p, q) of two-sex mixing functions and any time $t \geq 0$, nonnegative affinity functions $g(x, y)$ and $h(x, y)$ can be found so that*

$$\begin{aligned}
p(x, y) &= \bar{p}(y)\left[\frac{R_f(x)R_m(y)}{V_m} + g(x, y)\right], \\
q(y, x) &= \bar{q}(x)\left[\frac{R_f(x)R_m(y)}{V_f} + h(y, x)\right],
\end{aligned} \tag{2.22}$$

where the affinities satisfy

$$g(x, y) = h(y, x) + R_f(x)R_m(y)\left[\frac{1}{V_f} - \frac{1}{V_m}\right]. \tag{2.23}$$

Any pair (p, q) satisfying the axioms (A1)–(A3) *is in the form above.*

Including affinities in the marriage function is an overarching idea in the work of many authors (see [112, 38, 84]). In the previous section the affinities were parameters in the marriage function independent of the densities of females and males. This is not the case within the axiomatic framework discussed in this section, where the affinities depend implicitly on single females and single males through (2.23). Although there are two affinities, the equality (2.23) guarantees that the marriage function is a one-parameter family of operators (as far as affinities are concerned). In other words, if we know the affinities for females as well as the parameters (2.21), the affinities for males can be determined from (2.23) and vice versa.

Estimating the parameters in the Castillo-Chávez model has also been a topic of discussion. Castillo-Chávez and Hsu Schmitz worked on the problem in the context of sexual mixing [50, 52, 51]. The problem of inferring the mixing probabilities (respectively, the preferences) from data has been addressed by Blythe, Castillo-Chávez, and Casella [6, 16], and by Pugliese [108].

2.7 Nonhomogeneous marriage models

The most controversial of the properties of the marriage function is the homogeneity property (P3). This property is widely accepted and steadily present in the demographic literature. Almost all of the works on the two-sex problem (the only exceptions that we have encountered are discussed below) assume homogeneity in one form or another. The main reason is that homogeneity ensures scale invariance and helps avoid pathologies. More precisely, we would like to have (approximately) the same per capita marriage rate for each state in the U.S. as for the whole country. This effect is achieved if these marriage rates are scale invariant or, in mathematical terms, homogeneous of degree zero. Another possible reason why the homogeneity condition is so rarely challenged is that a homogeneous nonlinearity of degree one is a "natural generalization" of the linear case, and it can be hoped to preserve some of the important properties of the linear model and, in particular, the exponential growth [69].

However, some authors believe that the homogeneity assumption is too restrictive an approximation and can and should be avoided. A typical argument against it is that it does not hold at low population densities because the time to find an appropriate mate increases significantly. Moreover, some techniques for rigorous derivation of the marriage function lead to functions that are not homogeneous but rather asymptotically homogeneous (that is, homogeneous for large values of the densities). In particular, this occurs when the mechanism of encounters is refined by introducing two time-scales to describe the formation of married couples (on a slow scale) as the final product of several encounters occurring on a fast scale.

This modeling approach was adopted by Hsu and Fredrickson [49] and subsequently, in the context of ordinary differential equations models, by Heesterbeek and Metz [46]. A rough description of the argument can be obtained by introducing a new "intermediate" variable $k(x, y, t)$ denoting the age density of the nonstable (courting) couples occurring on a fast time-scale. Then, equation (2.8) of the Fredrickson–Hoppensteadt model is modified into

$$\begin{cases} \dfrac{\partial c}{\partial t} + \dfrac{\partial c}{\partial x} + \dfrac{\partial c}{\partial y} = -\sigma(x, y, t)c(x, y, t) \\ \qquad\qquad\qquad\qquad + \rho(x, y)k(x, y, t), & x, y, t > 0, \\ c(x, 0, t) = c(0, y, t) = 0, & x, y \geq 0, t > 0, \\ c(x, y, 0) = c^0(x, y), & x, y \geq 0, \end{cases} \tag{2.24}$$

where it is assumed that nonstable couples turn into married (stable) couples at a rate of $\rho(x, y)$ on a slow scale. For its part, the density $k(x, y, t)$ satisfies the equation

$$\frac{\partial k}{\partial t} + \frac{\partial k}{\partial x} + \frac{\partial k}{\partial y} = -(\mu_f(x) + \mu_m(y))k(x, y, t) - \delta_k k(x, y, t)$$
$$+ m_k \left[s_f(x) - \int_0^\omega k(x, y, t)dy \right] \left[s_m(y) - \int_0^\omega k(x, y, t)dx \right], \tag{2.25}$$

where the mating process is described by a mass-action law. The rates δ_k and m_k in (2.25) are much larger than the others so that, at the slow scale, this equation can be approximated by

$$\delta_k k(x, y) = m_k \left[s_f(x) - \int_0^\omega k(x, y)dy \right] \left[s_m(y) - \int_0^\omega k(x, y)dx \right], \tag{2.26}$$

thus providing a form for $k(x, y)$, as a function of $s_f(\cdot)$ and $s_m(\cdot)$, to be used in (2.24). In essence this leads to an expression for the marriage function. However, in order to provide an explicit form for the marriage function by the argument above, one has to solve (2.26). In [49] the authors provide a particular solution under the assumption that $K = m_k/\delta_k$ does not depend on the ages x and y. This leads to the marriage function

$$\mathcal{M}(x, y, t; s_f, s_m) = \frac{2\rho(x, y)K s_f(x) s_m(y)}{1 + K(u + v) + \sqrt{1 + 2K(u + v) + K(u - v)^2}}, \tag{E11}$$

where $u = \int_0^\omega s_f(x, t)\, dx$, $v = \int_0^\omega s_m(y, t)\, dy$.

In their article [46] Heesterbeek and Metz consider the general problem of deriving contact rates for marriage and epidemic models, and they provide a rigorous treatment based on singular perturbation techniques. The function they obtain in the context of nonstructured models is similar to the function proposed by Hsu and Fredrickson.

Though the marriage functions thus obtained do not satisfy the homogeneity property, they approximate homogeneous functions for large population densities, and, therefore, in the case of population growth, they may be considered not to contradict the homogeneity assumption. Since deterministic population models—such as the ones considered in this book—are generally valid only for large population densities, the homogeneity assumption on the marriage function in this case is quite appropriate and we shall use it throughout.

2.8 Coale and McNeil's risk of first marriage model

In the early 1970s Coale himself, and jointly with McNeil, discovered that "there are precisely defined age patterns of nuptiality, readily approximated by a simple mathematical expression, that are followed very closely indeed in populations under widely different social conditions" [21, 23]. In particular, they remark that—despite the cultural variety in age of first marriage and proportions of individuals never marrying in various societies—the proportion of women who have married for the first time by a given age among all women in the cohort who will ever marry seems to have a standard form in different populations, differing only in origin, area, and horizontal scale. This proportion can be treated as a probability distribution function of age, F. It turns out that, even though there is no simple formula to express F in closed form, if the horizontal scale is chosen so that $x = 0$ represents $xmin$ ("the earliest age at which a consequential number of first marriages occur" [23]), there is indeed a simple formula for the *risk of first marriage* defined by

$$r(x) = \frac{F'(x)}{[1 - F(x)]}.$$

In fact an excellent fit of Swedish data from the 1860s is obtained using the double exponential curve

$$r(x) = 0.174\, e^{-4.411\, e^{-0.309x}}. \tag{2.27}$$

We should note that, if we denote by $g = g(s)$ the probability density function for F, we have

$$F(x) = \int_0^x g(s)\, ds = 1 - e^{-\int_0^x r(t)\, dt}, \tag{2.28}$$

and we readily see that g and r are related by the simple formula

$$g(x) = r(x)\, e^{-\int_0^s r(t)\, dt}.$$

In Figure 2.1 we show the function $F(x)$ numerically computed from (2.27) using (2.28).

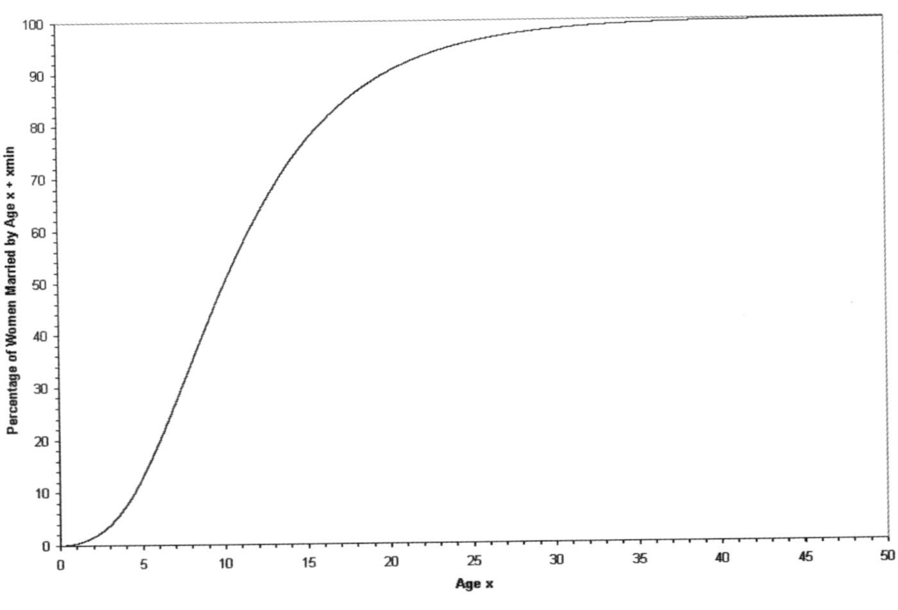

Figure 2.1. *The function $F(x)$ as computed from* (2.27).

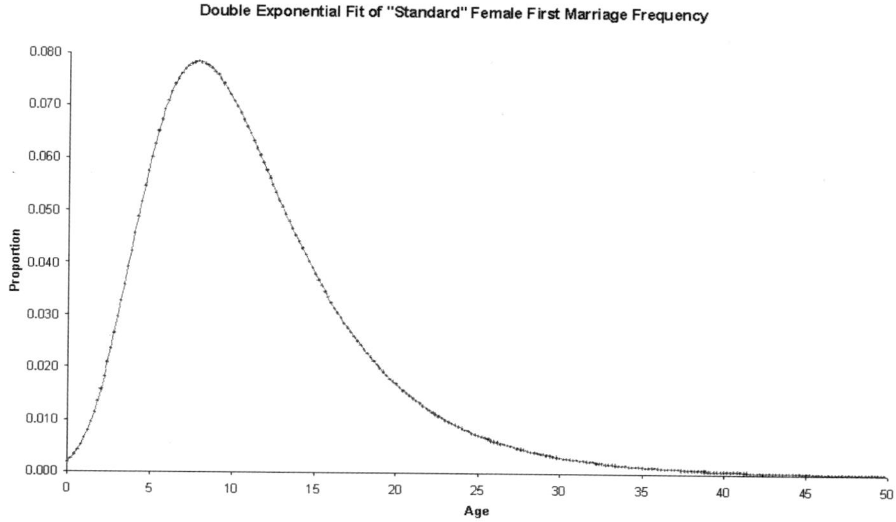

Figure 2.2. *The function $\bar{g}(x)$ fitting empirical data.*

Now, the empirical standard first marriage frequency, based on data for Swedish women during the period 1865–1869, actually leads to the approximation for $g(x)$ (see Figure 2.2)

$$\bar{g}(x) = 0.1946\, e^{-0.174(x-6.062)-e^{-0.2881(x-6.062)}},$$

which is similar to a frequency function of a Gompertz extreme value distribution.

The absolute value of the area between the standard curve and $\bar{g}$ is only 1.6% [23]. (We note here that in this article the vertical scale of the graph comparing the Swedish data with $\bar{g}$ on page 745 is mistakenly shown 10 times bigger than it should be.)

The values of the probability density function for F obtained from this r differ from those of $\bar{g}$ by less than 0.001 for any x, as seen in Figure 2.3, depicting the difference $\bar{g} - g$.

Coale and McNeil also show that the general form of the approximation $\bar{g}(x)$ to the probability density $g(x)$ is

$$\bar{g}(x) = \frac{\lambda}{\Gamma(\alpha/\lambda)}\, e^{-\alpha(x-\mu)-e^{-\lambda(x-\mu)}},$$

where $\mu = a + (1/\lambda)\psi(\alpha/\lambda)$, with $\psi = \Gamma'/\Gamma$, $\lambda = 0.2881$, and $\alpha = 0.174$; Γ is the usual gamma function and $a = 11.36$ is the mean of the standard fertility schedule. This function $\bar{g}$ happens to be extremely close to the convolution of a normal distribution of attainment of marriageable age and three exponentially distributed delays. These delays have means of $\frac{1}{0.174+2\times 0.2881} = 1.33$ years, $\frac{1}{0.174+0.2881} = 2.16$ years, and $\frac{1}{0.174} = 5.75$ years. The mean of the normal distribution, 2.12, is equal to 11.36 minus the sum of these three delays, and the standard deviation is 2.0 years. Actually, its variance equals the variance of $\bar{g}$ minus the sum of the variances of the three exponentially distributed delays.

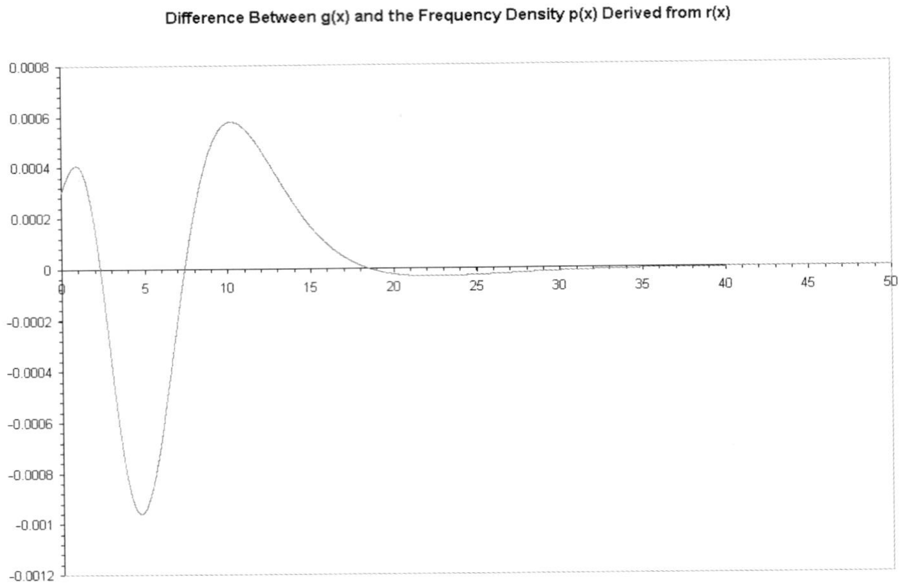

Figure 2.3. *The difference between* $g(x)$ *and* $\bar{g}(x)$.

Coale and McNeil go on to suggest that, since the distribution of first marriages is closely approximated by the convolution of the normal distribution of ages of entry into marriageability and three exponentially distributed delays, it is natural to ask whether these delays correspond to identifiable events that precede marriage and follow entry into marriageable age. They suggest that

> "In contemporary populations of Western European origin, in which marriage is typically a ceremony that unites a couple who select each other on the basis of mutual preference, we may conjecture that the age of becoming marriageable is the age at which serious dating, or going steady, begins; that the longest delay is the time between becoming marriageable and meeting (or starting to keep frequent company with) the eventual husband; and that the two shorter delays are the period between beginning to date the future husband and engagement, and between engagement and marriage."

Under this proviso, it follows that the distribution of duration of acquaintance before marriage has a density that is the convolution of the densities of two exponential distributions $(\alpha + \lambda)\, e^{-(\alpha+\lambda)x}$ and $(\alpha + 2\lambda)\, e^{-(\alpha+2\lambda)x}$, that is,

$$\hat{F}(x) = 1 + \left(\frac{\alpha + \lambda}{\lambda}\right) e^{-(\alpha+2\lambda)x} - \left(\frac{\alpha + 2\lambda}{\lambda}\right) e^{-(\alpha+\lambda)x}.$$

The authors then compare the predictions based on the standard curves with data from a study of French couples published in 1964, with a distribution of ages at first marriage having a mean of 23.1 years and a standard deviation of 4.60 years. They compute the three delays as 0.93, 1.52, and 4.02 years, giving a mean age of reaching marriageability of 16.6 years and a mean interval from acquaintance to marriage of 2.45 years. They observe a remarkable agreement between the percentages of couples with less than six months, one year, two years, and three years of acquaintance before marriage and those predicted using $\hat{F}$.

Coale and McNeil also apply this model to the distribution of the ages of females at first marriage in the United States in the 1960s. The normal distribution of age of becoming marriageable that they obtain has a mean of 15.6 years and a standard deviation of 1.52 years. The three delays they compute for this population are 10.5, 17, and 45.5 months, with a sum just above six years—giving a mean age at marriage slightly under 22 years.

The observed patterns of marriage should be reflected in any marriage function and, in principle, could be used both as a test of the marriage function and as a source for parameter identification.

Chapter 3

Well-Posedness of the Fredrickson–Hoppensteadt Two-Sex Model

Nella vita del signor Palomar c'è stata un'epoca in cui la sua regola era questa: primo, costruire nella sua mente un modello, il più perfetto, logico, geometrico possibile; secondo, verificare se il modello s'adatta ai casi pratici osservabili nell'esperienza; terzo, apportare le correzioni necessarie perchè modello e realtà coincidano. ... Il modello è per definizione quello in cui non c'è niente da cambiare, quello che funziona alla perfezione; mentre la realtà vediamo bene che non funziona e che si spappola da tutte le parti; dunque non resta che costringerla a prendere la forma del modello, con le buone o con le cattive.[1]

—I. Calvino, "Il modello dei modelli" from
Mr. Palomar, 1983

The Fredrickson–Hoppensteadt model described in the previous chapter will actually be our "model of models," the one "most perfect, logical, and geometrically possible," according to Mr. Palomar's first rule of thumb.

Motivated by our choice, we now continue with the investigation of the mathematical properties of the model. In fact in this chapter we establish that the Fredrickson–Hoppensteadt model is well posed in the sense that, given a set of initial conditions, it has a unique nonnegative solution. We also establish that, under appropriate conditions, the solution is differentiable and satisfies the equations in the classical sense. Knowledge of the mathematical aspects of the model will enable us to set up numerical methods, which we will present in Chapter 4. In particular, to prove convergence of the finite difference

[1]In the life of Mr. Palomar there was a time in which his rule of thumb was this: first, to build in his mind a model, the most perfect, logical, and geometrically possible; second, to verify if the model adapts to practical situations observed in experience; third, to provide the necessary corrections for model and reality to coincide.... A model is, by definition, that in which there is nothing to change, that works perfectly, while we well see that reality does not work and crashes everywhere. Thus, one can only force it to take the shape of the model, whether it likes it or not.

algorithm used as a basis for the simulator, we assume certain differentiability properties of the solution that we shall prove in section 3.3 (see also [1]).

The simulations that we shall perform in Chapter 4 will show that the Fredrickson–Hoppensteadt model can be successfully used for projecting human populations—at least for short intervals of time. However, we are aware of the fact that "reality crashes everywhere" and that further modeling will be necessary to reach a better description.

Interest in age-structured two-sex models is relatively recent, although the models date from the 1970s and 1980s. The early analysis of such models is mostly restricted to special, simpler cases [38]. Continuous and classical solutions are discussed by the present authors in [83]. Most of the results on the well-posedness of age-structured two-sex population models are concerned with existence and uniqueness of solutions in spaces of integrable functions. Inaba [59] considers a two-sex model for human reproduction by first marriage—that is, newborns are produced only in first-marriage couples—and establishes the well-posedness of the model and the existence of persistent solutions. Similar results are contained in a published version of the above-mentioned article [60]. The well-posedness of Hadeler's model is obtained in [85] as a result of the application of integrated semigroups.

3.1 An integral formulation of the model

The question of well-posedness of the model in the classical sense is interesting both from a theoretical point of view and also because it justifies the use of numerical methods. We begin the discussion by establishing existence and uniqueness of continuous solutions of the Fredrickson–Hoppensteadt model. What makes considering continuous solutions important, in light of population models, is that their existence can be established under biologically relevant conditions and very mild restrictions on the parameters. Moreover, they are in some sense "natural solutions" as they automatically possess a directional derivative in the direction of the characteristic lines and satisfy the model with this directional derivative in place of the partial derivatives.

We begin by introducing some biologically relevant quantities that will often be used throughout the rest of this work. We set

$$\begin{aligned}\pi_f(x,t;z) &= e^{-\int_0^z \mu_f(x-\tau,t-\tau)\,d\tau},\\ \pi_m(y,t;z) &= e^{-\int_0^z \mu_m(y-\tau,t-\tau)\,d\tau},\\ \pi_c(x,y,t;z) &= e^{-\int_0^z \sigma(x-\tau,y-\tau,t-\tau)\,d\tau}.\end{aligned} \tag{3.1}$$

These functions have a clear physical meaning: $\pi_f(x,t;z)$ is the probability that a female of age $x-z$ at time $t-z$ will survive to age x at time t (given that she has survived to age $x-z$). $\pi_m(y,t;z)$ has a similar meaning and, in addition, $\pi_c(x,y,t;z)$ is the probability that a couple for which, at time $(t-z)$ the wife is of age $(x-z)$ and the husband is of age $(y-z)$, will still be married at time t. In the special case when the sex-specific mortality rates and the splitting rate for the couples are time independent we will denote (3.1), respectively, as $\pi_f(x;z)$, $\pi_m(y;z)$, and $\pi_c(x,y;z)$. We also have

$$\pi_f(x;x) = \pi_f(x) = e^{-\int_0^x \mu_f(\sigma)d\sigma}$$

and

$$\pi_m(y;\, y) = \pi_m(y) = e^{-\int_0^y \mu_m(\sigma)d\sigma},$$

that is, we have the *female (respectively, male) survival probability* to age x (respectively, y) that expresses the probability at birth for a female (respectively, male) to survive to age x (respectively, y). These quantities are defined exactly as in the one-sex case (see (1.6)).

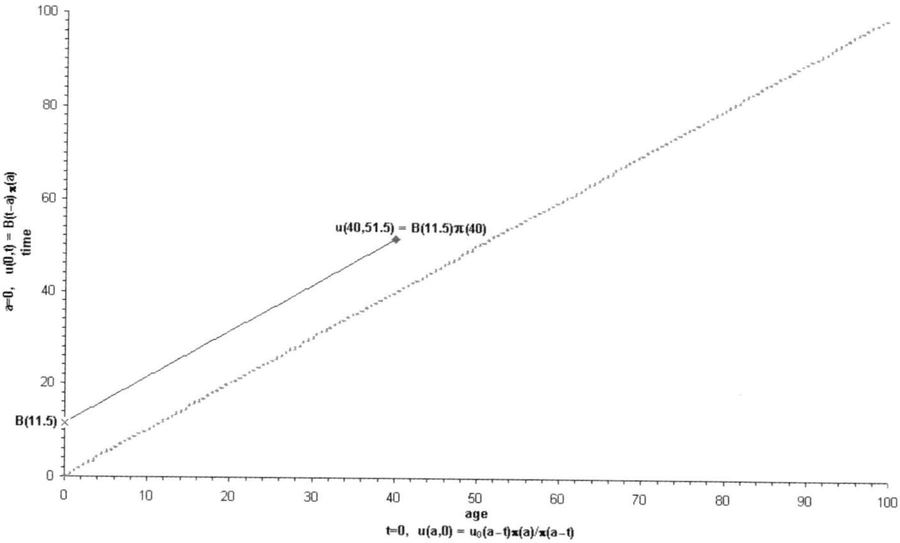

Figure 3.1. *The integration along characteristics.*

Because differential operators are difficult to work with, a typical approach toward the well-posedness of a differential equation problem is to rewrite it in integral form. Hence, we introduce an integral version of (2.6)–(2.8). To obtain this formulation we integrate the differential equations along the characteristic lines. The characteristics for (2.6) and (2.7) are the lines $x - t =$constant and $y - t =$constant, respectively (see Figure 3.1). We first integrate (2.6) and (2.7) and, similarly to the one-sex case, we obtain

$$u_f(x,t) = \begin{cases} u_f^0(x-t)\pi_f(x,t;\,t) & \text{for} \quad x \geq t, \\ B_f(t-x)\pi_f(x,t;\,x) & \text{for} \quad x < t, \end{cases}$$
$$u_m(y,t) = \begin{cases} u_m^0(y-t)\pi_m(y,t;\,t) & \text{for} \quad y \geq t, \\ B_m(t-y)\pi_m(y,t;\,x) & \text{for} \quad y < t. \end{cases} \tag{3.2}$$

Note that $B_f(t)$ and $B_m(t)$ depend on $c(x, y, t)$ through (2.6) and (2.7), respectively. We substitute (3.2) into (2.10) to obtain

$$
s_f(x,t) = \begin{cases} u_f^0(x-t)\pi_f(x,t;t) - \int_0^\omega c(x,y,t)\,dy & \text{for} \quad x \geq t, \\ B_f(t-x)\pi_f(x,t;x) - \int_0^\omega c(x,y,t)\,dy & \text{for} \quad x < t, \end{cases}
$$

$$
s_m(y,t) = \begin{cases} u_m^0(y-t)\pi_m(y,t;t) - \int_0^\omega c(x,y,t)\,dx & \text{for} \quad y \geq t, \\ B_m(t-y)\pi_m(y,t;x) - \int_0^\omega c(x,y,t)\,dx & \text{for} \quad y < t. \end{cases} \tag{3.3}
$$

Concerning (2.8), if we assume s_f and s_m as known, it turns into a linear ordinary differential equation along the lines $x - t =$ constant, $y - t =$ constant, and it can then be solved explicitly to give the following integral equation for the couples:

$$
c(x,y,t) = \begin{cases} c^0(x-t,y-t)\pi_c(x,y,t;t) \\ \qquad + \int_0^t \pi_c(x,y,t;\tau)\mathcal{M}(x-\tau,y-\tau,t-\tau;s_f,s_m)d\tau \\ \qquad\qquad \text{for} \quad x \geq t,\ y \geq t, \\ \int_0^x \pi_c(x,y,t;\tau)\mathcal{M}(x-\tau,y-\tau,t-\tau;s_f,s_m)d\tau \\ \qquad\qquad \text{for} \quad y \geq x,\ t > x, \\ \int_0^y \pi_c(x,y,t;\tau)\mathcal{M}(x-\tau,y-\tau,t-\tau;s_f,s_m)d\tau \\ \qquad\qquad \text{for} \quad x > y,\ t > y. \end{cases} \tag{3.4}
$$

Clearly, if (u_f, u_m, c) is a solution of the system (2.6)–(2.10), it also satisfies (3.2)–(3.4) above.

To show existence of a solution of the system of integral equations (3.2), (3.3), and (3.4) with densities of singles determined from (3.3) we shall iteratively define a sequence of functions $u_f^{(n)}$, $u_m^{(n)}$, and $c^{(n)}$ that we show converges to the solution of the integral equations (3.2), (3.4). We notice that the right-hand sides of (3.2) and (3.4) will give nonnegative left-hand sides for any choice of u_f, u_m, and c; however, the right-hand side of (3.3) will not always produce nonnegative values for the densities of single females and single males. This presents a technical difficulty in the proof that we overcome by transforming (3.2)–(3.4) into an equivalent problem. Namely we rewrite (3.2) as

$$u_f(x,t) = \begin{cases} u_f^0(x-t)\pi_f(x,t;t)e^{-t} \\ \qquad + \displaystyle\int_0^t e^{-\tau}\pi_f(x,t;\tau)u_f(x-\tau,t-\tau)\,d\tau \quad \text{for} \quad x \geq t, \\ B_f(t-x)\pi_f(x,t;x)e^{-x} \\ \qquad + \displaystyle\int_0^x e^{-\tau}\pi_f(x,t;\tau)u_f(x-\tau,t-\tau)\,d\tau \quad \text{for} \quad x < t, \end{cases}$$

$$u_m(y,t) = \begin{cases} u_m^0(y-t)\pi_m(y,t;t)e^{-t} \\ \qquad + \displaystyle\int_0^t e^{-\tau}\pi_m(y,t;\tau)u_m(y-\tau,t-\tau)\,d\tau \quad \text{for} \quad y \geq t, \\ B_m(t-y)\pi_m(y,t;x)e^{-y} \\ \qquad + \displaystyle\int_0^y e^{-\tau}\pi_m(y,t;\tau)u_m(y-\tau,t-\tau)\,d\tau \quad \text{for} \quad y < t, \end{cases} \tag{3.5}$$

and we change (3.4) in a similar way:

$$c(x,y,t) = \begin{cases} c^0(x-t,y-t)\pi_c(x,y,t;t)e^{-t} \\ \qquad + \displaystyle\int_0^t e^{-\tau}\pi_c(x,y,t;\tau)c(x-\tau,y-\tau,t-\tau)\,d\tau \\ \qquad + \displaystyle\int_0^t e^{-\tau}\pi_c(x,y,t;\tau)\mathcal{M}(x-\tau,y-\tau,t-\tau;s_f,s_m)d\tau \\ \qquad\qquad\qquad \text{for} \quad x \geq t,\ y \geq t, \\ \displaystyle\int_0^x e^{-\tau}\pi_c(x,y,t;\tau)c(x-\tau,y-\tau,t-\tau)\,d\tau \\ \qquad + \displaystyle\int_0^x e^{-\tau}\pi_c(x,y,t;\tau)\mathcal{M}(x-\tau,y-\tau,t-\tau;s_f,s_m)d\tau \\ \qquad\qquad\qquad \text{for} \quad y \geq x,\ t > x, \\ \displaystyle\int_0^y e^{-\tau}\pi_c(x,y,t;\tau)c(x-\tau,y-\tau,t-\tau)\,d\tau \\ \qquad + \displaystyle\int_0^y e^{-\tau}\pi_c(x,y,t;\tau)\mathcal{M}(x-\tau,y-\tau,t-\tau;s_f,s_m)d\tau \\ \qquad\qquad\qquad \text{for} \quad x > y,\ t > y, \end{cases} \tag{3.6}$$

where s_f and s_m are defined through (3.3).

In the next section we prove that the system of integral equations (3.5)–(3.6) has a continuous solution. Since the densities of females, males, and couples are then only continuous, they could, in principle, lack partial derivatives. However, it turns out that they have classical derivatives in the direction of the characteristics and they satisfy equations analogous to (2.6)–(2.8) with a directional derivative in place of the sum of the partial derivatives.

3.2 Existence and uniqueness of a continuous solution

We begin by introducing the mathematical setup as well as our assumptions. Let $T > 0$. Throughout this chapter we shall assume that $t \in [0, T]$. We make the following assumptions on the parameters of the model.

(H1) The female and male birth rates $\beta_f(x, y, t)$ and $\beta_m(x, y, t)$ are nonnegative and continuous on $Q \equiv [0, \omega] \times [0, \omega] \times [0, T]$, so they are bounded on Q and we set

$$\bar{\beta}_f = \sup_{(x,y,t)\in Q} |\beta_f(x, y, t)|, \quad \bar{\beta}_m = \sup_{(x,y,t)\in Q} |\beta_m(x, y, t)|.$$

(H2) The female and male mortality rates $\mu_f(x, t)$ and $\mu_m(y, t)$ and the dissolution rate for couples $\sigma(x, y, t)$ are nonnegative and continuous on the sets $[0, \omega) \times [0, T]$ and $[0, \omega) \times [0, \omega) \times [0, T]$, respectively. Moreover, they satisfy

$$\int_0^\omega \mu_f(x, t)\, dx = \infty, \quad \int_0^\omega \mu_m(y, t)\, dy = \infty$$

for every fixed t.

This condition is necessary to guarantee that the probability of survival of a female or male individual to the maximal age is zero.

(H3) The initial densities for females, males, and couples $u_f^0(x)$, $u_m^0(y)$, and $c^0(x, y)$ are nonnegative, continuous, and satisfy

$$\int_0^\omega c^0(x, y) dy \le u_f^0(x), \quad \int_0^\omega c^0(x, y) dx \le u_m^0(y).$$

These two conditions are necessary to guarantee that the solutions are nonnegative. More precisely, they imply that s_f^0 and s_m^0 are nonnegative. Thus, they guarantee that the initial conditions are physically meaningful. In particular, the first condition says that the density of all women of age x who are married to a man of any age is no larger than that of all women of age x (since the latter also include single women). The second condition has a similar meaning with the roles of the sexes reversed.

(H4) We assume that the initial and boundary conditions are compatible:

$$u_f^0(0) = B_f(0) = \int_0^\omega \int_0^\omega \beta_f(x, y, 0) c^0(x, y)\, dxdy,$$

$$u_m^0(0) = B_m(0) := \int_0^\omega \int_0^\omega \beta_m(x, y, 0) c^0(x, y)\, dxdy,$$

$$c^0(x, 0) = c^0(0, y) = 0 \qquad \text{for} \qquad x \ge 0,\ y \ge 0.$$

This requirement must be imposed because, in order to have continuity of the solution across the lines $x = t$ and $y = t$, we need the initial and boundary conditions to agree at the origin. The hypothesis is technical and is not usually satisfied by data.

(H5) The marriage function $\mathcal{M}(x, y, t; s_f, s_m)$ satisfies at least properties (P1), (P2), and (P4). Moreover, it is of the first kind,

$$\mathcal{M}(x, y, t; s_f, s_m) = \mathcal{F}\left[x, y, t, s_f(x, t), s_m(y, t)\right], \tag{3.7}$$

where $\mathcal{F}[x, y, t, f, m]$ is continuous on $[0, \omega]^2 \times [0, \infty)^3$ and is uniformly Lipschitz continuous in the variables f and m, i.e., there exists a constant $L > 0$ such that

$$\left|\mathcal{F}[x, y, t, f, m] - \mathcal{F}\left[x, y, t, \bar{f}, \bar{m}\right]\right| \leq L\left(|f - \bar{f}| + |m - \bar{m}|\right) \tag{3.8}$$

for $(x, y, t) \in [0, \omega]^2 \times [0, T]$.

Concerning this last condition (H5) on the marriage function, we remark that we have only assumed properties (P1), (P2), and (P4), which are needed in the proof of Theorem 3.1 below. We also note that among the sample marriage functions of the first kind (E1)–(E7) presented in Chapter 2, only (E4) does not satisfy the Lipschitz condition (3.8). We will further discuss this point in section 3.4, where we will also consider extensions to second and third kind marriage functions.

Next, we state and establish our main result in this section.

Theorem 3.1. *Under the assumptions* (H1)–(H5)*, the system of integral equations* (3.5)–(3.6)*, with the coupling relations* (3.3)*, has a unique, continuous, nonnegative solution.*

Proof. We iteratively define a sequence of functions $u_f^{(n)}, u_m^{(n)}, c^{(n)}$ as follows. We start by setting

$$u_f^{(0)}(x, t) \equiv 0, \quad u_m^{(0)}(y, t) \equiv 0, \quad c^{(0)}(x, y, t) \equiv 0. \tag{3.9}$$

Then, at each iteration, from $u_f^{(n)}(x, t)$, $u_m^{(n)}(y, t)$, $c^{(n)}(x, y, t)$, we first compute

$$c^{(n+1)}(x, y, t) = \begin{cases} c^0(x - t, y - t)\pi_c(x, y, t; t)e^{-t} \\ \quad + \displaystyle\int_0^t e^{-\tau}\pi_c(x, y, t; \tau)c^{(n)}(x - \tau, y - \tau, t - \tau)\, d\tau \\ \quad + \displaystyle\int_0^t e^{-\tau}\pi_c(x, y, t; \tau)\mathcal{M}(x - \tau, y - \tau, t - \tau; s_f^{(n)}, s_m^{(n)})d\tau \\ \qquad\qquad \text{for} \quad x \geq t, y \geq t, \\ \displaystyle\int_0^x e^{-\tau}\pi_c(x, y, t; \tau)c^{(n)}(x - \tau, y - \tau, t - \tau)\, d\tau \\ \quad + \displaystyle\int_0^x e^{-\tau}\pi_c(x, y, t; \tau)\mathcal{M}(x - \tau, y - \tau, t - \tau; s_f^{(n)}, s_m^{(n)})d\tau \\ \qquad\qquad \text{for} \quad y \geq x, t > x, \\ \displaystyle\int_0^y e^{-\tau}\pi_c(x, y, t; \tau)c^{(n)}(x - \tau, y - \tau, t - \tau)\, d\tau \\ \quad + \displaystyle\int_0^y e^{-\tau}\pi_c(x, y, t; \tau)\mathcal{M}(x - \tau, y - \tau, t - \tau; s_f^{(n)}, s_m^{(n)})d\tau \\ \qquad\qquad \text{for} \quad x > y, t > y, \end{cases} \tag{3.10}$$

where the densities of single female and single male individuals are determined from

$$
\begin{aligned}
s_f^{(n)}(x,t) &= u_f^{(n)}(x,t) - \int_0^\omega c^{(n)}(x,y,t)\,dy, \\
s_m^{(n)}(y,t) &= u_m^{(n)}(y,t) - \int_0^\omega c^{(n)}(x,y,t)\,dx.
\end{aligned}
\tag{3.11}
$$

Then, we compute

$$
u_f^{(n+1)}(x,t) = \begin{cases} u_f^0(x-t)\pi_f(x,t;t)e^{-t} \\ \quad + \displaystyle\int_0^t e^{-\tau}\pi_f(x,t;\tau)u_f^{(n)}(x-\tau,t-\tau)\,d\tau & \text{for} \quad x \geq t, \\ B_f^{(n+1)}(t-x)\pi_f(x,t;x)e^{-x} \\ \quad + \displaystyle\int_0^x e^{-\tau}\pi_f(x,t;\tau)u_f^{(n)}(x-\tau,t-\tau)\,d\tau & \text{for} \quad x < t, \end{cases}
$$

$$
u_m^{(n+1)}(y,t) = \begin{cases} u_m^0(y-t)\pi_m(y,t;t)e^{-t} \\ \quad + \displaystyle\int_0^t e^{-\tau}\pi_m(y,t;\tau)u_m^{(n)}(y-\tau,t-\tau)\,d\tau & \text{for} \quad y \geq t, \\ B_m^{(n+1)}(t-y)\pi_m(y,t;x)e^{-y} \\ \quad + \displaystyle\int_0^y e^{-\tau}\pi_m(y,t;\tau)u_m^{(n)}(y-\tau,t-\tau)\,d\tau & \text{for} \quad y < t, \end{cases}
\tag{3.12}
$$

where

$$
B_f^{(n+1)}(t) = \int_0^\omega \int_0^\omega \beta_f(x,y,t)c^{(n+1)}(x,y,t)\,dxdy \tag{3.13}
$$

and

$$
B_m^{(n+1)}(t) = \int_0^\omega \int_0^\omega \beta_m(x,y,t)c^{(n+1)}(x,y,t)\,dxdy. \tag{3.14}
$$

First, we establish that the iteration defined in (3.10)–(3.12) produces continuous nonnegative functions. In particular, we need s_f and s_m to be nonnegative in order to compute $\mathcal{M}(x,y,t;s_f^{(n)},s_m^{(n)})$. In order to see that, we first note that $u_f^{(0)}, u_m^{(0)}, c^{(0)}, s_f^{(0)}, s_m^{(0)}$, defined in (3.9), are obviously all continuous and nonnegative. In the next step we establish that, if $u_f^{(n)}, u_m^{(n)}, c^{(n)}, s_f^{(n)}$, and $s_m^{(n)}$ are continuous and nonnegative, then so are $u_f^{(n+1)}, u_m^{(n+1)}$, $c^{(n+1)}, s_f^{(n+1)}$, and $s_m^{(n+1)}$. Indeed, considering $c^{(n+1)}(x,y,t)$, we see that each piece in its definition in (3.10) is continuous. Discontinuities may only occur on the boundary planes of the region which divide each part of the space, that is, the half-planes $\{x = t, y \geq t\}$ and $\{x \geq t, y = t\}$. However, we see that, if $x = t$ (equivalently, if $y = t$), then the very first term in (3.10) contains as a multiplier $c^0(0, y-t)$ (respectively, $c^0(x-t, 0)$), which is zero. Moreover, the integrals defining the first piece of the function are exactly the same as the corresponding integrals defining the second piece, and also the integrals in the third piece. Consequently, $c^{(n+1)}(x, y, t)$ is continuous and, by its definition, it is nonnegative. Concerning the continuity of $u_f^{(n+1)}$ and $u_m^{(n+1)}$, we notice that each piece in the definition is continuous thanks to the assumptions and the continuity of the nth iterates. In addition, when $x = t$ (respectively, $y = t$), the integrals in the first and second pieces are the same and hypothesis (H4) implies that the first terms are equal, and we notice that

$$B_f^{(n+1)}(0) = \int_0^\omega \int_0^\omega \beta_f(x, y, 0)c^{(n+1)}(x, y, 0)dxdy$$
$$= \int_0^\omega \int_0^\omega \beta_f(x, y, 0)c^0(x, y)dxdy,$$

$$B_m^{(n+1)}(0) = \int_0^\omega \int_0^\omega \beta_m(x, y, 0)c^{(n+1)}(x, y, 0)dxdy$$
$$= \int_0^\omega \int_0^\omega \beta_m(x, y, 0)c^0(x, y)dxdy.$$

Finally, we note that, as a consequence of the previous considerations, $s_f^{(n+1)}$ and $s_m^{(n+1)}$ are also continuous. To prove that they are nonnegative we show that

$$\int_0^\omega c^{(n+1)}(x, y, t)dy \leq u_f^{(n+1)}(x, t).$$

Indeed, from the expression for $c^{(n+1)}$, if $x \geq t$, we obtain the following formula, after changing the order of integration:

$$\int_0^\omega c^{(n+1)}(x, y, t)dy$$

$$= \int_0^\omega c^0(x - t, y - t)\pi_c(x, y, t; t)e^{-t}dy$$

$$+ \int_0^t \int_\tau^\omega e^{-\tau}\pi_c(x, y, t; \tau)c^{(n)}(x - \tau, y - \tau, t - \tau)dyd\tau$$

$$+ \int_0^t \int_\tau^\omega e^{-\tau}\pi_c(x, y, t; \tau)\mathcal{M}(x - \tau, y - \tau, t - \tau; s_f^{(n)}, s_m^{(n)})dyd\tau.$$

An important step in this argument is noticing that

$$\begin{aligned} \pi_c(x, y, t; z) &\leq \pi_f(x, t; z), \\ \pi_c(x, y, t; z) &\leq \pi_m(y, t; z). \end{aligned} \tag{3.15}$$

These inequalities agree with our intuition since, e.g., the first one says that the probability of survival of a couple from age $x - t$ of the wife and age $y - t$ of the husband to ages x and y, respectively, is smaller than the probability of survival of the wife from age $x - t$ to age x—since a marriage can dissolve not only because the wife dies but also because the husband dies or the couple divorces.

Using (3.15) we have the estimate

$$\begin{aligned}
&\int_0^\omega c^{(n+1)}(x,y,t)dy \\
&\leq \pi_f(x,t;t)e^{-t}\int_0^\omega c^0(x-t,y)dy \\
&\quad + \int_0^t e^{-\tau}\pi_f(x,t;\tau)\int_0^\omega c^{(n)}(x-t,y,t-\tau)dyd\tau \\
&\quad + \int_0^t e^{-\tau}\pi_f(x,t;\tau)\int_0^\omega \mathcal{M}(x-\tau,y,t-\tau;s_f^{(n)},s_m^{(n)})dyd\tau \\
&\leq \pi_f(x,t;t)e^{-t}u_f^0(x-t) \\
&\quad + \int_0^t e^{-\tau}\pi_f(x,t;\tau)\int_0^\omega c^{(n)}(x-t,y,t-\tau)dyd\tau \\
&\quad + \int_0^t e^{-\tau}\pi_f(x,t;\tau)s_f^{(n)}(x-\tau,t-\tau)d\tau,
\end{aligned}$$

where we have used property (P4) of the marriage function. From this inequality, (3.11), and (3.12) we have

$$\begin{aligned}
\int_0^\omega c^{(n+1)}(x,y,t)dy &\leq \pi_f(x,t;t)e^{-t}u_f^0(x-t) \\
&\quad + \int_0^t e^{-\tau}\pi_f(x,t;\tau)u_f^{(n)}(x-\tau,t-\tau)d\tau = u_f^{(n+1)}(x,t)
\end{aligned}$$

for $x \geq t$. A similar argument holds for $x < t$ and also for proving that $s_m^{(n+1)}$ is nonnegative.

The next step is to establish that the sequences $u_f^{(n)}$, $u_m^{(n)}$, and $c^{(n)}$ converge, respectively, to some functions u_f, u_m, and c, which will clearly satisfy (3.2) and (3.4). We can easily verify the latter if we take the limit as n goes to infinity in the equalities (3.10) and (3.12).

To see that the sequences converge we introduce for $n \geq 1$ the notation:

$$\begin{aligned}
F_n(t) &= \sup_{x\in[0,\omega]} |u_f^{(n)}(x,t) - u_f^{(n-1)}(x,t)|, \\
M_n(t) &= \sup_{x\in[0,\omega]} |u_m^{(n)}(y,t) - u_m^{(n-1)}(y,t)|, \\
C_n(t) &= \sup_{x,y\in[0,\omega]} |c^{(n)}(x,y,t) - c^{(n-1)}(x,y,t)|,
\end{aligned} \tag{3.16}$$

and

$$N_n(t) = M_n(t) + F_n(t) + C_n(t). \tag{3.17}$$

Note that $M_n(t)$, $F_n(t)$, $C_n(t)$, and $N_n(t)$ are nonnegative functions of t. In addition, we set

$$F_0 = \sup_{x\in[0,\omega]} |u_f^0(x)|, \quad M_0 = \sup_{x\in[0,\omega]} |u_m^{(0)}(y)|, \quad C_0 = \sup_{x,y\in[0,\omega]} |c^{(0)}(x,y)|.$$

Now we note that by the choice of the initial step (3.9) we have

$$c^{(1)}(x,y,t) = \begin{cases} c^0(x-t,y-t)\pi_c(x,y,t;t)e^{-t} & \text{for} \quad x \geq t, y \geq t, \\ 0 & \text{otherwise,} \end{cases}$$

and

$$u_f^{(1)}(x,t) = \begin{cases} u_f^0(x-t)\pi_f(x,t;t)e^{-t} & \text{for} \quad x \geq t, \\ B_f^{(1)}(t-x)\pi_f(x,t;x)e^{-x} & \text{for} \quad x < t, \end{cases}$$

$$u_m^{(1)}(y,t) = \begin{cases} u_m^0(y-t)\pi_m(y,t;t)e^{-t} & \text{for} \quad y \geq t, \\ B_m^{(1)}(t-y)\pi_m(y,t;x)e^{-y} & \text{for} \quad y < t. \end{cases}$$

Thus, it is easy to see that

$$|u_f^{(1)}(x,t) - u_f^{(0)}(x,t)| \leq K_f(F_0 + C_0) \leq K_f N_0,$$

$$|u_m^{(1)}(y,t) - u_m^{(0)}(y,t)| \leq K_m(M_0 + C_0) \leq K_m N_0,$$

$$|c^{(1)}(x,y,t) - c^{(0)}(x,y,t)| \leq K_c C_0 \leq K_c N_0,$$

where $N_0 = (M_0 + F_0 + C_0)$ and K_f, K_m, and K_c are constants determined below. Consequently,

$$N_1(t) \leq K N_0,$$

where $K = K_f + K_m + K_c$.

We now use the Lipschitz condition provided by assumption (H5) to have, for $n \geq 1$,

$$|\mathcal{M}(x,y,t;s_f^{(n)},s_m^{(n)}) - \mathcal{M}(x,y,t;s_f^{(n-1)},s_m^{(n-1)})|$$

$$\leq L\left(|s_f^{(n)}(x,t) - s_f^{(n-1)}(x,t)| + |s_m^{(n)}(y,t) - s_m^{(n-1)}(y,t)|\right)$$

$$\leq L\left(F_n(t) + M_n(t) + 2\omega C_n(t)\right).$$

Hence, from (3.10) and (3.12) we have

$$F_{n+1}(t) \leq K_1 C_{n+1}(t) + \int_0^t F_n(\tau)\,d\tau,$$

$$M_{n+1}(t) \leq K_2 C_{n+1}(t) + \int_0^t M_n(\tau)\,d\tau, \tag{3.18}$$

$$C_{n+1}(t) \leq K_3 \int_0^t C_n(\tau)\,d\tau + L\int_0^t F_n(\tau)\,d\tau + L\int_0^t M_n(\tau)\,d\tau,$$

where $K_1 = \bar{\beta}_f\omega^2$, $K_2 = \bar{\beta}_m\omega^2$, and $K_3 = 1 + 2\omega L$. Substituting the third inequality above into the first and second, we obtain

$$F_{n+1}(t) \leq K_1K_3\int_0^t C_n(\tau)\,d\tau + (1+K_1L)\int_0^t F_n(\tau)\,d\tau + K_1L\int_0^t M_n(\tau)\,d\tau,$$

$$M_{n+1}(t) \leq K_2K_3\int_0^t C_n(\tau)\,d\tau + K_2L\int_0^t F_n(\tau)\,d\tau + (1+K_2L)\int_0^t M_n(\tau)\,d\tau,$$

$$C_{n+1}(t) \leq K_3\int_0^t C_n(\tau)\,d\tau + L\int_0^t F_n(\tau)\,d\tau + L\int_0^t M_n(\tau)\,d\tau.$$

Letting $K_f = \max\{K_1K_3, 1 + K_1L\}$, $K_m = \max\{K_2K_3, 1 + K_2L\}$, and $K_c = \max\{K_3, L\}$, the inequalities above can be written as

$$F_{n+1}(t) \le K_f\left[\int_0^t C_n(\tau)\,d\tau + \int_0^t F_n(\tau)\,d\tau + \int_0^t M_n(\tau)\,d\tau\right],$$

$$M_{n+1}(t) \le K_m\left[\int_0^t C_n(\tau)\,d\tau + \int_0^t F_n(\tau)\,d\tau + \int_0^t M_n(\tau)\,d\tau\right],$$

$$C_{n+1}(t) \le K_c\left[\int_0^t C_n(\tau)\,d\tau + \int_0^t F_n(\tau)\,d\tau + \int_0^t M_n(\tau)\,d\tau\right].$$

Summing these three inequalities we get

$$N_{n+1}(t) \le K\int_0^t N_n(\tau)\,d\tau, \tag{3.19}$$

where the constant K is defined above. We recall that we also have $N_1(t) \le KN_0$. By induction it follows that

$$N_n(t) \le N_0\frac{(K)^n t^{(n-1)}}{(n-1)!} \le N_0\frac{(K)^n T^{(n-1)}}{(n-1)!}.$$

This says that the sequences $u_f^{(n)}(x,t)$, $u_m^{(n)}(y,t)$, and $c^{(n)}(x,y,t)$ generated by (3.9)–(3.14) are uniformly convergent. In fact we have

$$\sup_{x\in[0,\omega]}\left|u_f^{n+m}(x,t) - u_f^n(x,t)\right| \le \sum_{i=n}^{m+n} N_i(t) \le \sum_{i=n}^{\infty}\frac{K^iT^{i-1}}{(i-1)!} \to 0 \quad \text{as} \quad n\to\infty$$

and the sequence $u_f^{(n)}$ is a Cauchy sequence uniformly on $[0,\omega]\times[0,T]$. The same result holds for $u_m^{(n)}$ and $c^{(n)}$; hence, there exist $u_f(x,t)$, $u_m(y,t)$, and $c(x,y,t)$, nonnegative and continuous limits of these sequences. This concludes the proof of existence.

To prove uniqueness we assume that there is a second set of functions $(\bar{u}_f, \bar{u}_m, \bar{c})$ that also satisfies (3.5) and (3.6). Then, if we set

$$\begin{aligned}
\Delta_f(t) &= \sup_{x\in[0,\omega]}|u_f(x,t) - \bar{u}_f(x,t)|,\\
\Delta_m(t) &= \sup_{x\in[0,\omega]}|u_m(y,t) - \bar{u}_m(y,t)|,\\
\Delta_c(t) &= \sup_{x,y\in[0,\omega]}|c(x,y,t) - \bar{c}(x,y,t)|,
\end{aligned}$$

we obtain, as before,

$$\Delta_f(t) + \Delta_m(t) + \Delta_c(t) \le K\int_0^t \left[\Delta_f(\tau) + \Delta_m(\tau) + \Delta_c(\tau)\right]d\tau,$$

and Gronwall's lemma implies that $\Delta_f(t) + \Delta_m(t) + \Delta_c(t) \equiv 0$. Therefore,

$$u_f(x,t) \equiv \bar{u}_f(x,t), \quad u_m(y,t) \equiv \bar{u}_m(y,t), \quad c(x,y,t) \equiv \bar{c}(x,y,t),$$

and the uniqueness is proved. □

Now, we would like to know in what sense the solution of the system of integral equations (3.5)–(3.6) solves the original system of differential equations (2.6)–(2.8). In fact the continuous solutions $u_f(x,t)$, $u_m(y,t)$, and $c(x,y,t)$ that we have found (and, consequently, continuous $s_f(x,t)$, $s_m(y,t)$), as we prove in the proposition below, have some differentiability properties that give a particular meaning to the way they solve the original equations. First, we have the following result.

Proposition 3.2. *If* $u_f(x,t)$, $u_m(y,t)$, *and* $c(x,y,t)$ *are continuous solutions of* (3.5) *and* (3.6), *then the first two have directional derivatives in the direction of the vector* $(1,1)$ *and the last in the direction of* $(1,1,1)$, *satisfying the equations*

$$\begin{aligned}
&\lim_{h\to 0}\frac{1}{h}[u_f(x+h,t+h)-u_f(x,t)] = -\mu_f(x,t)u_f(x,t),\\
&\lim_{h\to 0}\frac{1}{h}\,[u_m(y+h,t+h)-u_m(y,t)] = -\mu_m(y,t)u_m(y,t),\\
&\lim_{h\to 0}\frac{1}{h}\,[c(x+h,y+h,t+h)-c(x,y,t)]\\
&\qquad\qquad = -\sigma(x,y,t)c(x,y,t)+\mathcal{M}(x,y,t;s_f,s_m).
\end{aligned} \tag{3.20}$$

Proof. To show the first relation in (3.20) we differentiate (3.5) along the characteristic lines. Assume x,t fixed and $x>t$ and consider $u_f(x+h,t+h)$. Recall that the probability of survival of a female has the meaning

$$\pi_f(x,t;t) = e^{-\int_0^t \mu_f(x-t+\sigma,\sigma)\,d\sigma}$$

and

$$\pi_f(x,t;t-\tau) = e^{-\int_\tau^t \mu_f(x-t+\sigma,\sigma)\,d\sigma}.$$

Hence, for $x>t$ we have from (3.2)

$$\begin{aligned}
u_f(x+h,t+h) &= u_f^0(x-t)e^{-(t+h)}e^{-\int_0^{t+h}\mu_f(x-t+\sigma,\sigma)\,d\sigma}\\
&\quad+\int_0^{t+h} e^{-(t+h-\tau)}e^{-\int_\tau^{t+h}\mu_f(x-t+\sigma,\sigma)\,d\sigma}u_f(x-t+\tau,\tau)\,d\tau.
\end{aligned}$$

Consequently, differentiating this last relation with respect to h and setting $h=0$, we obtain the first equality in (3.20).

Analogously, we can establish the result for $x\le t$ and for $u_m(y,t)$ and $c(x,y,t)$, so that the proof is complete. $\square$

We note that, as a consequence of the previous proposition, and using similar arguments, it is possible to show that $s_f(x,t)$ and $s_m(y,t)$ also have continuous directional derivatives along the lines with slope one. We provide this in the proposition below.

Proposition 3.3. *The functions $s_f(x,t)$ and $s_m(y,t)$, given by*

$$s_f(x,t) = u_f(x,t) - \int_0^\omega c(x,y,t)\,dy,$$

$$s_m(y,t) = u_m(y,t) - \int_0^\omega c(x,y,t)\,dx,$$

have continuous directional derivatives in the direction of the vector $(1,1)$.

Proof. Since we have shown this result for $u_f(x,t)$ and $u_m(x,t)$, it will be sufficient to establish that

$$\int_0^\omega c(x,y,t)\,dx \qquad \text{and} \qquad \int_0^\omega c(x,y,t)\,dy$$

have continuous directional derivatives in the direction of the vector $(1,1)$. We shall prove this for one of the integrals—for the other one the argument is analogous. Performing the integration of $c(x,y,t)$ with respect to y, from (3.6) we have

$$u_f(x,t) - s_f(x,t) = \int_0^\omega c(x,y,t)\,dy$$

$$= \begin{cases} \displaystyle\int_t^\omega c^0(x-t,y-t)\pi_c(x,y,t;t)e^{-t}\,dy \\ \displaystyle\quad + \int_0^t\int_\tau^\omega e^{-\tau}\pi_c(x,y,t;\tau)c(x-\tau,y-\tau,t-\tau)\,dyd\tau \\ \displaystyle\quad + \int_0^t\int_\tau^\omega e^{-\tau}\pi_c(x,y,t;\tau)\mathcal{M}(x-\tau,y-\tau,t-\tau;s_f,s_m)\,dyd\tau \\ \hfill \text{for} \quad x \geq t, \\ \displaystyle\int_0^x\int_\tau^\omega e^{-\tau}\pi_c(x,y,t;\tau)c(x-\tau,y-\tau,t-\tau)\,dyd\tau \\ \displaystyle\quad + \int_0^x\int_\tau^\omega e^{-\tau}\pi_c(x,y,t;\tau)\mathcal{M}(x-\tau,y-\tau,t-\tau;s_f,s_m)\,dyd\tau \\ \hfill \text{for} \quad t > x. \end{cases} \tag{3.21}$$

Then, we can proceed as in the previous proof. In fact, if we change the variable of integration in the first term for the case $x \geq t$, we obtain

$$\int_0^{\omega-t} e^{-t}c^0(x-t,\gamma)\pi_c(x,\gamma+t,t;t)\,d\gamma,$$

where π_c can be conveniently written as

$$\pi_c(x,\gamma+t,t;t) = e^{-\int_0^t \sigma(x-t+\eta,\gamma+\eta,\eta)\,d\eta}.$$

Moreover, after several changes of variables, the second term can be written in the form

$$\int_0^t\int_\tau^\omega e^{-\tau}\pi_c(x,y,t;\tau)c(x-\tau,y-\tau,t-\tau)\,dyd\tau$$
$$= \int_0^t\int_{t-\tau}^\omega e^{-(t-\tau)}\pi_c(x,\eta+t-\tau,t;t-\tau)c(x-t+\tau,\eta,\tau)\,d\eta d\tau$$

and

$$\pi_c(x, \eta + t - \tau, t; t - \tau) = e^{-\int_\tau^t \sigma(x-t+\alpha, \eta-\tau+\alpha, \alpha)\, d\alpha}.$$

A similar change of variables transforms the third integral into

$$\int_0^t \int_{t-\tau}^{\omega} e^{-(t-\tau)} \pi_c(x, \eta + t - \tau, t; t - \tau) \mathcal{M}(x - t + \tau, \eta, \tau; s_f, s_m)\, d\eta\, d\tau.$$

We note that, in the integrands of all these expressions, c and $\mathcal{M}$ depend on x and t only through the expression $(x - t)$. Thus, for small $\eta > 0$, they are both constant on the line segment $(x + h, \xi, t + h)$, $-\eta < h < \eta$. Therefore, it is evident that, for $x \geq t$, the limit

$$\lim_{h \to 0} \frac{1}{h} \left[\int_0^{\omega} c(x + h, y, t + h) dy - \int_0^{\omega} c(x, y, t) dy \right]$$

does exist. The case $x < t$ can be treated similarly since the only difference is that the upper limits of the integrals are x instead of t. This concludes the proof. □

Proposition 3.3 gives a regularity result for the solution of the problem that, in fact, is quite natural and expected because of the nature of the problem. However, further conditions are necessary in order to have a solution that has continuous partial derivatives and strictly satisfies the differential problem (2.6)–(2.8). Existence of such a solution—called the *classical solution*—is discussed in the following section.

3.3 Conditions for existence of a classical solution

From the proof of Theorem 3.1 and from Propositions 3.2 and 3.3, it follows that the solution of the system of integral equations (3.5)–(3.6) not only is continuous but also has a continuous directional derivative in the direction of the vector $(1, 1)$ and the vector $(1, 1, 1)$, as appropriate. This means that, if the functions u_f and u_m have one continuous partial derivative, the other one also exists and is continuous. Denoting by D_t, D_x, and D_y, respectively, the partial derivatives with respect to t, x, and y, we summarize the above in the following proposition.

Proposition 3.4. *If $u_f(x, t)$ and $u_m(y, t)$ are continuous solutions of* (3.5)–(3.6) *and if $D_t u_f(x, t)$ and $D_t u_m(y, t)$ exist and are continuous, then $D_x u_f(x, t)$ and $D_y u_m(y, t)$ also exist and are continuous.*

Therefore, the task of proving continuity of partial derivatives is simplified because for the functions u_m and u_f to be continuously differentiable it only remains to be seen that one of their partial derivatives exists and is continuous. We shall show this for the time derivatives.

Concerning differentiability of the function $c(x, y, t)$, we note that establishing that the time derivative exists and is continuous does not automatically imply that c has all its derivatives continuous. However, it does imply that $\int_0^{\omega} c(x, y, t)dx$ and $\int_0^{\omega} c(x, y, t)dy$ have continuous derivatives, which, in turn, leads to the fact that s_f and s_m have continuous partial derivatives by (3.3). By (3.4) it then follows that c is continuously differentiable in

x, y, and t because the right-hand side is, provided the marriage function is continuously differentiable. Thus, in order to establish continuous differentiability of the solution, it is enough to establish the existence of a continuous time derivative.

Before we continue with the statement and the proof of this fact, we need to introduce some additional hypotheses.

(H6) The initial densities for females and males $u_f^0(\cdot)$ and $u_m^0(\cdot)$ are continuously differentiable on the entire age interval $[0, \omega]$.

(H7) The initial density of the couples $c^0(\cdot, \cdot)$ is continuously differentiable in both variables on the entire age interval $[0, \omega]$.

In addition, we assume that the partial derivatives with respect to either variable are zero as long as the age of the female in the couple or the age of the male in the couple is zero. In particular,

$$D_x c^0(x, y)|_{x=0} = D_x c^0(x, y)|_{y=0} = D_y c^0(x, y)|_{x=0} = D_y c^0(x, y)|_{y=0} = 0.$$

This condition is not really a restriction since there is a strictly positive minimal age below which individuals are not eligible for coupling.

(H8) The partial derivatives with respect to time of the birth rates for females and males $D_t \beta_f(x, t)$ and $D_t \beta_m(x, t)$ exist and are continuous on the set $[0, \omega] \times [0, \omega] \times [0, T]$.

Similarly, the partial derivatives with respect to time of the female and male mortality rates $D_t \mu_m(x, t)$ and $D_t \mu_f(y, t)$ are continuous on $[0, \omega) \times [0, T]$.

Moreover, we assume that the separation rate for couples $\sigma(x, y, t)$ is continuously differentiable on $[0, \omega] \times [0, \omega] \times [0, T]$.

Finally, we assume boundedness of the time derivatives of the corresponding probabilities of survival, namely, we assume that

$$D_t \pi_m(y, t; z), \quad D_t \pi_f(x, t; z), \quad D_t \pi_c(x, y, t; z)$$

are bounded on $[0, \omega] \times [0, T]$ and $[0, \omega] \times [0, \omega] \times [0, T]$, as appropriate.

We note that the mortality rates, including the separation rate for couples, are not even defined at the endpoints of the age interval so that we cannot require boundedness of their derivatives.

(H9) The basic function $\mathcal{F}[x, y, t, f, m]$ defining the marriage function of first kind in assumption (H5) is differentiable with respect to the variables t, f, and m, and there exists $L > 0$ such that

$$\begin{aligned} &\left|D_t \mathcal{F}[x, y, t, f, m] - D_t \mathcal{F}[x, y, t, \bar{f}, \bar{m}]\right| \\ &\quad + \left|D_f \mathcal{F}[x, y, t, f, m] - D_f \mathcal{F}[x, y, t, \bar{f}, \bar{m}]\right| \\ &\quad\quad + \left|D_m \mathcal{F}[x, y, t, f, m] - D_m \mathcal{F}[x, y, t, \bar{f}, \bar{m}]\right| \\ &\quad\quad\quad \leq L\left(|f - \bar{f}| + |m - \bar{m}|\right). \end{aligned} \tag{3.22}$$

(H10) The initial data for each sex satisfy the following first order compatibility conditions, that is, compatibility conditions for the derivatives:

$$-\frac{d}{dx}u_f^0(0) - \mu_f(0,0)u_f^0(0) = \frac{d}{dt}B_f^0(0),$$

$$-\frac{d}{dy}u_m^0(0) - \mu_m(0,0)u_m^0(0) = \frac{d}{dt}B_m^0(0).$$

A few remarks on these assumptions are now in order.

First, notice that, in assumption (H8), except for $\sigma(x,y,t)$, the vital parameters of the model are only assumed to have continuous partial derivatives with respect to t. Hence, in the special case when they are time independent, their continuity is sufficient for the existence of classical solutions.

In the same assumption we include the boundedness of the time derivative of the survival probabilities. This is a condition satisfied for a large family of mortality rates. However, it is not trivial, that is, there are mortality rates for which it is not satisfied, as suggested by the example

$$\mu_f(x,t) = \frac{1}{(\omega - x)} + \frac{1}{(t+\alpha)},$$

where α is some positive constant. We note that this function satisfies some preliminary properties that mortality rates should have, e.g., it grows unboundedly as x approaches ω and decreases with time, which reflects the fact that with time the probability of survival to a given age increases. The probability of survival that corresponds to the mortality rate above is given by

$$\pi_f(x,t;z) = \frac{(\omega - x)(t - z + \alpha)}{(t+\alpha)(\omega - x + z)}.$$

It is easy to see that this expression is bounded by one. The boundedness of the time derivative of $\pi_f(x,t;x)$ is also easy to see. However, the time derivative of $\pi_f(x,t;t)$ contains a term

$$\frac{\alpha(\omega - x)}{(t+\alpha)(\omega - x + t)^2},$$

which is unbounded when $t = 0$ and x approaches ω. The boundedness of the derivatives of the survival probability has been found to play an important role in numerically approximating the one-sex model [56].

Another important remark concerns the assumption in (H9). Actually, this assumption implies that, if $s_f(x,t)$ and $s_m(y,t)$ are continuously differentiable with respect to t, then the marriage function $\mathcal{M}(x,y,t;s_f,s_m)$ is continuously differentiable with respect to t. In fact we have

$$\begin{aligned} D_t\mathcal{M}(x,y,t;s_f,s_m) &= D_t\mathcal{F}[x,y,t,s_f(x,t),s_m(y,t)] \\ &\quad + D_f\mathcal{F}[x,y,t,s_f(x,t),s_m(y,t)]D_ts_f(x,t) \\ &\quad + D_m\mathcal{F}[x,y,t,s_f(x,t),s_m(y,t)]D_ts_m(y,t). \end{aligned} \tag{3.23}$$

This property is a basic ingredient in the proof of Theorem 3.6 below, but it is a very serious one because (H9) is only satisfied by the female dominance function (E1) and the male dominance function (E2). The main reason is the fact that the marriage function is assumed homogeneous of degree one and homogeneous functions are not differentiable at zero unless they are linear. This point will be considered further in section 3.4 after we present the main results. There, we will discuss how these results can be extended to be valid for the harmonic mean marriage function (E3).

Finally, we note that the compatibility condition (H10) is the complement of assumption (H4), which is needed to have continuity of the solution. It is similar to the one assumed by Gurtin and MacCamy in [37] for the one-sex model.

In order to prove regularity of the solutions, we will use the approximating sequence $u_f^{(n)}(x,t)$, $u_m^{(n)}(y,t)$, $c^{(n)}(x,y,t)$ that we defined in the previous section. Since we start with $u_f^{(0)}(x,t)$, $u_m^{(0)}(y,t)$, $c^{(0)}(x,y,t)$—obviously continuously differentiable—we find that the iterates are continuously differentiable with respect to t. In fact, if we assume that $u_f^{(n)}(x,t)$, $u_m^{(n)}(y,t)$, $c^{(n)}(x,y,t)$ are continuously differentiable, then noticing that $s_f^{(n)}(x,t)$ and $s_m^{(n)}(y,t)$ given by (3.11) are also continuously differentiable with respect to t, we can differentiate the right-hand side of (3.10). Once the differentiability of $c^{(n+1)}$ is established, we can pass it to (3.13)–(3.14) and, finally, to (3.12). When performing the derivatives with respect to time, we see that the expressions are continuous across the line $x = t$ (respectively, $y = t$) as a result of the consistency conditions (H4) and (H10). Hypotheses (H1)–(H3) and (H6)–(H8) then guarantee that these functions are continuous on each of the half-planes $x > t$ and $x < t$ (respectively, $y > t$ and $y < t$). Hypotheses (H1)–(H10) also imply the continuity of $D_t c^n$ in $\mathbb{R}^3_+$.

A second step consists of proving the following lemma.

Lemma 3.5. *The sequence of partial derivatives is uniformly bounded, that is, there exists a constant C independent of n such that for $n \geq 0$*

$$|D_t u_f^{(n)}(x,t)| + |D_t u_m^{(n)}(y,t)| + |D_t c^{(n)}(x,y,t)| \leq C. \tag{3.24}$$

Proof. To prove this lemma we first note that $D_t c^{(n+1)}$ has the form

$$\begin{aligned} D_t c^{(n+1)}(x,y,t) = {} & \mathcal{K}_1(x,y,t) \\ & + \mathcal{K}_3(x,y,t;t)\mathcal{M}(x-t, y-t, 0; s_f^{(n)}, s_m^{(n)}) \\ & + \int_0^{x\wedge y\wedge t} \mathcal{K}_2(x,y,t;\tau) c^{(n)}(x-\tau, y-\tau, t-\tau)\,d\tau \\ & + \int_0^{x\wedge y\wedge t} \mathcal{K}_2(x,y,t;\tau)\mathcal{M}(x-\tau, y-\tau, t-\tau; s_f^{(n)}, s_m^{(n)})\,d\tau \\ & + \int_0^{x\wedge y\wedge t} \mathcal{K}_3(x,y,t;\tau) D_t c^{(n)}(x-\tau, y-\tau, t-\tau)\,d\tau \\ & + \int_0^{x\wedge y\wedge t} \mathcal{K}_3(x,y,t;\tau) D_t \mathcal{M}(x-\tau, y-\tau, t-\tau; s_f^{(n)}, s_m^{(n)})\,d\tau, \end{aligned} \tag{3.25}$$

where $\mathcal{K}_i(x, y, t; \tau)$ are continuous functions thanks to assumptions (H1)–(H10), as already remarked. Similar expressions hold for $D_t u_f^{(n)}$ and $D_t u_m^{(n)}$. Then, if we set

$$\tilde{F}_n(t) = \sup_{x\in[0,\omega]} |D_t u_f^{(n)}(x,t)|, \quad \tilde{M}_n(t) = \sup_{y\in[0,\omega]} |D_t u_m^{(n)}(y,t)|,$$

$$\tilde{C}_n(t) = \sup_{x,y\in[0,\omega]} |D_t c^{(n)}(x,y,t)|,$$

and

$$\tilde{N}_n(t) = \tilde{F}_n(t) + \tilde{M}_n(t) + \tilde{C}_n(t),$$

from (3.25) we have

$$\tilde{C}_{n+1}(t) \le K_1 + K_2 \int_0^t \tilde{N}_n(s)ds,$$

where K_1 and K_2 are appropriate constants. To derive this estimate we have used the fact that the sequences $s_f^{(n)}$, $s_m^{(n)}$, and $c^{(n)}$ are uniformly bounded so that from (3.23) and (H9) we have

$$|D_t\mathcal{M}(x, y, t; s_f^{(n)}, s_m^{(n)})| \le H_1 + H_2(|D_t s_f^{(n)}(x,t)| + |D_t s_m^{(n)}(y,t)|)$$

for appropriate constants H_1 and H_2. Then, using this inequality in the equations for $D_t u_f^{(n+1)}$ and $D_t u_m^{(n+1)}$ we see that

$$\begin{aligned}\tilde{F}_{n+1}(t) &\le K_3 + K_4 \int_0^t \tilde{F}_n(\tau)d\tau + K_5 \int_0^t \tilde{C}_n(\tau)d\tau,\\ \tilde{M}_{n+1}(t) &\le K_3 + K_4 \int_0^t \tilde{M}_n(\tau)d\tau + K_5 \int_0^t \tilde{C}_n(\tau)d\tau,\end{aligned} \tag{3.26}$$

for appropriate constants K_3, K_4, and K_5. Thus, we have

$$\tilde{N}_{n+1}(t) \le K + K\int_0^t \tilde{N}_n(\tau)d\tau,$$

where K is a new constant depending on $K_1, \dots, K_5$. Next, using induction on n, starting with $\tilde{N}_0(t) \equiv 0$, we see that

$$\tilde{N}_n(t) \le Ke^{Kt} \qquad \text{for} \quad n \ge 0.$$

Finally, (3.24) follows directly with $C = Ke^{KT}$. □

Now we are ready to prove the main result in this section.

Theorem 3.6. *Assume* (H1)–(H10) *hold. Then, the sequences of continuous functions* $D_t u_f^n(x,t)$, $D_t u_m^n(y,t)$, *and* $D_t c_f^n(x,y,t)$ *are uniformly convergent.*

Proof. To show that the sequences are uniformly convergent we adopt the notation

$$\begin{aligned}\widehat{F}_n(t) &= \sup_{x\in[0,\omega]} |D_t u_f^{(n)}(x,t) - D_t u_f^{(n-1)}(x,t)|,\\ \widehat{M}_n(t) &= \sup_{y\in[0,\omega]} |D_t u_m^{(n)}(y,t) - D_t u_m^{(n-1)}(y,t)|,\\ \widehat{C}_n(t) &= \sup_{x,y\in[0,\omega]} |D_t c^{(n)}(x,y,t) - D_t c^{(n-1)}(x,y,t)|,\end{aligned} \tag{3.27}$$

and we set

$$\widehat{N}_n(t) = \widehat{F}_n(t) + \widehat{M}_n(t) + \widehat{C}_n(t).$$

Now, by (3.23) and assumption (3.22) in (H9) we obtain

$$\begin{aligned}&\left|D_t\mathcal{M}(x,y,t;s_f^{(n+1)},s_m^{(n+1)}) - D_t\mathcal{M}(x,y,t;s_f^{(n)},s_m^{(n)})\right|\\ &\quad\le L\left(1+|D_t s_f^{(n)}(x,t)|+|D_t s_m^{(n)}(y,t)|\right)N_n(t)\\ &\qquad + H\left(|D_t s_f^{(n)}(x,t) - D_t s_f^{(n-1)}(x,t)| + |D_t s_m^{(n)}(y,t) - D_t s_m^{(n-1)}(y,t)|\right),\end{aligned}$$

where $N_n(t)$ has been defined in (3.17) and we have used the uniform boundedness of the sequences $s_f^{(n)}$ and $s_m^{(n)}$ to get the constant H. Then, by the previous lemma we obtain from (3.25) the estimate

$$\widehat{C}_{n+1}(t) \le H_1 \int_0^t \left(N_n(\tau) + \widehat{N}_n(\tau)\right) d\tau,$$

where H_1 is a new appropriate constant independent of n. This estimate can be used in the expressions for $D_t u_f^{(n+1)}$ and $D_t u_m^{(n+1)}$ so that we finally see

$$\widehat{N}_{n+1}(t) \le H_2 \int_0^t \left(N_n(\tau) + \widehat{N}_n(\tau)\right) d\tau.$$

If we now take into account the estimate (3.19), we get

$$N_{n+1}(t) + \widehat{N}_{n+1}(t) \le H_3 \int_0^t \left(N_n(\tau) + \widehat{N}_n(\tau)\right) d\tau$$

and, if A is a constant such that

$$N_1(t) + \widehat{N}_1(t) \le A,$$

we have

$$N_n(t) + \widehat{N}_n(t) \le A\frac{H_3^{n-1}t^{n-1}}{(n-1)!},$$

as can be seen by induction. As in the proof of Theorem 3.1, this last inequality implies uniform convergence of the sequences $D_t u_f^{(n)}$, $D_t u_m^{(n)}$, and $D_t c^{(n)}$. □

We summarize the results of this section in the following theorem.

Theorem 3.7. *Assume* (H1)–(H10) *hold. Then, there exists a unique continuously differentiable solution of the system of integral equations* (3.5)–(3.6), *i.e., a classical solution of* (2.6)–(2.8). *Moreover,* $s_f(x,t)$ *and* $s_m(y,t)$ *given by* (3.3) *are differentiable.*

The previous result is a reasonably satisfactory approach to the two-sex problems. The fact that the mathematical assumptions are somewhat strict and do not allow the use of most of the marriage functions is actually not an obstacle because we may use regular approximations of the marriage functions that fall within the framework of Theorem 3.7. In the following section we shall discuss these problems.

3.4 Extensions and developments

In this section we discuss some problems that arose in the previous sections in connection with Theorems 3.1 and 3.6.

First, we consider the problem of the Lipschitz continuity stated as condition (3.8) of (H5). In fact we have noticed that (3.8) is not satisfied by the first kind marriage function (E4), defined by

$$\mathcal{F}[x,y,t,f,m] = \rho(x,y)\sqrt{fm}.$$

In this specific case modeling arguments help to solve the problem. It has been observed that the geometric mean model fails at low values of the densities, so that a good description of the marriage process may be provided by modifying (E4) into the function

$$\mathcal{F}_\epsilon[x,y,t,f,m] = \rho(x,y)\begin{cases} \sqrt{fm} & \text{if} \quad fm \ge \epsilon^2, \\ \dfrac{1}{\epsilon} fm & \text{if} \quad fm \le \epsilon^2. \end{cases}$$

This new function still follows the geometric mean law except at low densities, but it satisfies (3.8) and is mathematically suitable.

A second point, still related to assumption (H5), concerns its extension to include the marriage functions of second and third kind, such as (E8), (E9), and (E10) presented in section 2.5. Indeed, these examples do not satisfy the following consequence of (3.8):

$$\begin{aligned} &\left|\mathcal{M}(x,y,t;s_f,s_m) - \mathcal{M}(x,y,t;\bar{s}_f,\bar{s}_m)\right| \\ &\qquad \le L\left(\left|s_f(x,t) - \bar{s}_f(x,t)\right| + \left|\bar{s}_m(y,t) - \bar{s}_m(y,t)\right|\right), \end{aligned}$$

which we used in the proof of Theorem 3.1, but they rather suggest that the right condition to impose is actually the following:

$$\begin{aligned} &\int_0^\omega \int_0^\omega \left|\mathcal{M}(x,y,t;s_f,s_m) - \mathcal{M}(x,y,t;\bar{s}_f,\bar{s}_m)\right| dxdy \\ &\qquad \le L\left(\int_0^\omega \left|s_f(x,t) - \bar{s}_f(x,t)\right| dx + \int_0^\omega \left|\bar{s}_m(y,t) - \bar{s}_m(y,t)\right| dy\right). \end{aligned}$$

This condition, satisfied by (E8), (E9), and (E10), allows the extension of Theorem 3.1 within the framework of functions that are integrable with respect to the age variables. The well-posedness theorem resulting within this context provides existence and uniqueness in a different class of functions. The idea of the proof of such a theorem is the same as that in section 3.2, but the iterates are now shown to converge in the L^1 norm, uniformly with respect to t. The differences in the proof are essentially technical and the outcome is a solution in a weaker sense. A discussion of well-posedness in this weaker setting can be found in [55] for this model and in [85] for Hadeler's.

Finally, we are concerned with the regularity hypotheses in Theorem 3.6, in particular with assumption (H9). In fact, as we noted before, the harmonic mean (E3) does not satisfy this hypothesis, so that further considerations are necessary. In order to fill this gap, we consider the following family of modified harmonic mean functions (for simplicity we assume they do not depend explicitly on t):

$$\mathcal{F}_\epsilon(x, y, f, m) = \rho(x, y)\frac{fm}{\epsilon + f + m}.$$

For every $\epsilon > 0$ these functions are Lipschitz continuous and their partial derivatives with respect to f and m satisfy condition (3.22). More precisely, we have

$$D_f\mathcal{F}_\epsilon(x, y, f, m) = \rho(x, y)\frac{\epsilon m + m^2}{(\epsilon + f + m)^2}.$$

Consequently, for any fixed $\epsilon > 0$ we have

$$|D_f\mathcal{F}_\epsilon(x, y, f, m) - D_f\mathcal{F}_\epsilon(x, y, \bar{f}, \bar{m})| \leq \frac{2\bar{\rho}}{\epsilon}|f - \bar{f}| + \frac{4\bar{\rho}}{\epsilon}|m - \bar{m}|,$$

where $\bar{\rho}$ is the supremum of the function $\rho(x, y)$. A similar inequality holds for $\mathcal{F}$, and condition (3.22) is satisfied.

The other hypotheses can also be verified and thus Theorem 3.7 implies that there exists a unique continuously differentiable solution of the system of integral equations (3.5)–(3.6) with the first kind marriage function $\mathcal{M}_\epsilon$ defined through $\mathcal{F}_\epsilon$.

Let us denote this solution with $(u_f^\epsilon, u_m^\epsilon, c^\epsilon)$. We can show that these sequences converge uniformly, respectively, to u_f, u_m, and c—the solutions of (3.5)–(3.6) with the marriage function given by the harmonic mean (E3).

This result is based on the observation that

$$\begin{aligned} &|\mathcal{M}_\epsilon(x, y, t; s_f^\epsilon, s_m^\epsilon) - \mathcal{M}(x, y, t; s_f, s_m)| \\ &\qquad \leq \epsilon + 2|s_f^\epsilon(x, t) - s_f(x, t)| + 2|s_m^\epsilon(y, t) - s_m(y, t)|. \end{aligned}$$

Applying a similar approach as on several previous occasions, we have

$$\begin{aligned} F_\epsilon(t) &= \sup_{x\in[0,\omega]} |u_f^\epsilon(x, t) - u_f(x, t)|, \\ M_\epsilon(t) &= \sup_{y\in[0,\omega]} |u_m^\epsilon(y, t) - u_m(y, t)|, \\ C_\epsilon(t) &= \sup_{x,y\in[0,\omega]} |c^\epsilon(x, y, t) - c(x, y, t)|, \end{aligned} \tag{3.28}$$

and, based on the inequality above, we derive the inequalities

$$\begin{aligned} F_\epsilon(t) &\leq \bar{\beta}_f \omega^2 C_\epsilon(t) + \int_0^t F_\epsilon(\tau) d\tau, \\ M_\epsilon(t) &\leq \bar{\beta}_m \omega^2 C_\epsilon(t) + \int_0^t M_\epsilon(\tau) d\tau, \\ C_\epsilon(t) &\leq \epsilon t + 2\int_0^t F_\epsilon(\tau) d\tau + 2\int_0^t M_\epsilon(\tau) d\tau + 5\int_0^t C_\epsilon(\tau) d\tau. \end{aligned} \tag{3.29}$$

Let $N_\epsilon(t) = F_\epsilon(t) + M_\epsilon(t) + C_\epsilon(t)$. Substituting the last inequality into the first two and summing the three inequalities, we obtain the relation

$$N_\epsilon(t) \leq K_1 \epsilon + K_2 \int_0^t N_\epsilon(\tau) d\tau,$$

where K_1 and K_2 are appropriate constants that depend on $\bar{\beta}_f$, $\bar{\beta}_m$, ω, and T, but not on ϵ. Gronwall's inequality then implies

$$N_\epsilon(t) \leq \epsilon \, K_1 e^{K_2 t},$$

from which uniform convergence follows.

The reason we have pointed out this convergence is that, since the numerical methods that we will discuss in Chapter 4 require regularity of the solution, we may first approximate our model with an approximating $\mathcal{F}_\epsilon$ and then discretize this modified model to provide numerical approximations. If ϵ is chosen sufficiently small (e.g., smaller than the machine zero of the computer used), the numerical approximations obtained for the modified model will be identical to those obtained for the original one with the harmonic mean marriage function (E3) that results from $\epsilon = 0$.

Chapter 4
Numerical Methods

– C'est bougrement idéaliste, dites donc, ce que vous me racontez là.
– Réaliste vous voulez dire: les nombres sont des réalités. Ils existent, les nombres! Ils existent autant que cette table, plus que cette table sempiternel exemple des philosophes, infiniment plus que cette table bang!
– Vous ne pourriez pas faire un peu moins de bruit, dit le garçon.[1]

—R. Queneau, *Odile*, 1937

Not only do numbers exist but they are also the only "réalités" by which the real world interacts with the models we may design. Thus, in order to use any of the models described in the previous chapters, we have to produce numbers from them, to compare with data, and to make projections about a population. Then, it is necessary to use numerical methods to approximate the solution of the model. This is due to the fact that—even though the theory proves that solutions exist and are unique—unfortunately there are no explicit formulas to express them in terms of the initial data and model parameters.

Among the many types of numerical methods that exist for the approximation of solutions of partial differential equations, the finite difference method of characteristics is particularly well suited for age-structured population models because of the simplicity of their characteristic curves. They are straight lines with unit slope, $a - t =$ constant.

A few papers can be found in the literature specifically addressing numerical methods for two-sex population models. The earliest one, dating back to 1988, is based on the finite element method and is suboptimal in its rate of convergence [90]. Approximately one year later another paper described a first order numerical method based on the finite difference method of characteristics combined with the trapezoidal rule for quadratures of the birth

[1]– Well! It is damned idealistic what you are telling me.
– Realistic you mean: numbers are realities. They exist, the numbers! They exist as much as this table, more than this table, eternal example of philosophers, infinitely more than this table bang!
– Couldn't you be a little less noisy, says the waiter.

integrals [1]. A second order method based on the finite difference method of characteristics combined with the trapezoidal rule was introduced in [91].

In this chapter we shall first be concerned with the description of a numerical method—essentially that of Arbogast and Milner [1]—that is based on the finite difference method of characteristics for the discretization of the differential equations, combined with quadrature rules for the integrals that describe the reproductive process of the population and the age densities of married females and males. Later, we shall use numerical simulations carried out with the method described to compare various marriage functions in their ability to project the total numbers of newborn females, newborn males, and couples.

4.1 Finite difference method of characteristics

We shall now describe a particular kind of numerical method that will be used later to make projections for the year Y_2 starting with data from the year $Y_1 < Y_2$.

Let $T = Y_2 - Y_1 > 0$ be the "final time," so that $t = 0$ corresponds to the year Y_1 and $t = T$ corresponds to the year Y_2. Let $N \in \mathbf{N}$ be the number of steps one wants the simulation to take in order to go from time $t = 0$ to time $t = T$.

The differential operator used to model the evolution of the age densities of females and males is

$$D = \frac{\partial}{\partial t} + \frac{\partial}{\partial a}, \tag{4.1}$$

where

$$\begin{aligned} Du(a,t) &= \frac{\partial u}{\partial t}(a,t) + \frac{\partial u}{\partial a}(a,t) \\ &= \lim_{h_1 \to 0} \frac{u(a,t+h_1) - u(a,t)}{h_1} + \lim_{h_2 \to 0} \frac{u(a+h_2,t) - u(a,t)}{h_2} \\ &= \lim_{h_1 \to 0} \frac{u(a+h_1,t+h_1) - u(a+h_1,t)}{h_1} + \lim_{h_2 \to 0} \frac{u(a+h_2,t) - u(a,t)}{h_2} \\ &= \lim_{h \to 0} \frac{u(a+h,t+h) - u(a,t)}{h}. \end{aligned}$$

Since this differential operator is of first order with constant coefficients, we discretize age and time using the same fixed discretization parameter, $h = T/N$, also called the *time-step*. Moreover, along a characteristic line $t = a + K$ (K being a constant), a partial differential equation of McKendrick–von Foerster type (1.4) satisfies the *ordinary differential equation*

$$U'(a) = Du(a, a+K) = -\mu(a)\, u(a, a+K) = -\mu(a)\, U(a),$$

where the function

$$U(a) = u(a, a+K).$$

This allows the approximation of the solution U by any numerical method for ordinary differential equations, as well as by using quadratures for the integral in the explicit formula

$$U(a) = U(0)\, e^{-\int_0^a \mu(s)\, ds}.$$

Finite difference methods for first order ordinary differential equations are based on the approximation of the limit of the difference quotients that defines the derivative by the difference quotients themselves. For example,

$$f'(x) = \lim_{h\to 0} \frac{f(x+h) - f(x)}{h} \approx \frac{f(x+h) - f(x)}{h}$$

is the basis for Euler's method.

Numerical methods for a partial differential equation—e.g., the McKendrick–von Foerster equation—that are based on a finite difference method for the corresponding ordinary differential equation on the characteristic curves are called finite difference methods of characteristics.

The numerical method we shall now describe and later use is based on a Crank–Nicolson type replacement of the derivative along characteristics, combined with the trapezoidal rule for the quadrature of the integrals that describe births and densities of married individuals of each sex. As we shall see in the next section, this leads to second order accuracy.

We introduce now some convenient notation to simplify the description of the numerical method. For i, j, n nonnegative integers let

$$x_i = ih, \quad y_j = jh, \quad \text{and} \quad t^n = nh.$$

For a function $f(x, y, t)$ of one or two age arguments and time (x, y, and/or t—and correspondingly i, j, and/or n—may be missing) let

$$f^n_{i,j} = f(x_i, y_j, t^n).$$

In the numerical approximations we do not actually use time-independent vital and demographic rates. In fact we take time-dependent rates that change linearly in time by interpolating the known (and different) rates at two different times. For example, for simulations starting in 1970, we use mortality data for the year 1970 to create the piecewise constant function $\mu^1(a)$ and we also use mortality data for the year 1980 to create the piecewise constant function $\mu^2(a)$. Using these we finally define a time-dependent mortality

$$\mu(a, t) = \mu^1(a) + t\,\frac{1}{10}\left[\mu^2(a) - \mu^1(a)\right].$$

We process natality rates, annulment/divorce rates, and marriage rates similarly.

Concerning the initial data, we shall assume that the age densities of females, males, and couples are given by compactly supported functions in the age interval $[0, M]$, that is,

$$u^0_f(x) = u^0_m(y) = c^0(x, y) = 0 \ \text{ for } \ x \geq M \ \text{ or } \ y \geq M. \tag{4.2}$$

Of course, for consistency reasons (4.2) must always be satisfied if $M \geq \omega$, where the probabilities of survival are null at age ω, that is, $\pi_g(\omega) = 0$ for $g = f, m$. Therefore, M should always be chosen no larger than ω. Then, *even in the case* $\omega = \infty$, it is still the case that all the age densities are compactly supported in age for any fixed time. In particular, for $0 \leq t \leq T$, it is verified that

$$u_f(x, t) = u_m(y, t) = c(x, y, t) = 0 \ \text{ for } \ x \geq M + t \ \text{ or } \ y \geq M + t. \tag{4.3}$$

Let now

$$L = M/h = \text{number of steps to move from age } a = 0 \text{ to age } a = M, \tag{4.4}$$

which we assume is a positive integer. Also, in order to simplify the numerical algorithm, we shall assume that all model parameters are bounded, except for mortality rates. Moreover, we shall assume that no individual marries at birth, so that

$$\mathcal{M}(0, y, t; s_f, s_m) = \mathcal{M}(x, 0, t; s_f, s_m) = 0. \tag{4.5}$$

Assumption (4.5) readily leads to the conclusion that, as $h \to 0$, the density of couples with either partner of age h is of order $O(h^2)$. For example, if $h \leq y$, then

$$\begin{aligned} c(h, y, t) &= c(0, y-h, t-h) + h\left(\frac{\partial c}{\partial t} + \frac{\partial c}{\partial x} + \frac{\partial c}{\partial y}\right)(0, y-h, t-h) + O(h^2) \\ &= h\,\mathcal{M}\big(0, y-h, t-h; s_f(0), s_m(y-h)\big) \\ &\quad - h\,\sigma(0, y-h)\,c(0, y-h, t-h) + O(h^2) \\ &= O(h^2), \end{aligned} \tag{4.6}$$

since $c(0, y-h, t-h) = \mathcal{M}\big(0, y-h, t-h; s_f(0), s_m(y-h)\big) = 0$.

For $0 \leq i, j \leq L+n$, $0 \leq n \leq N$, we shall define approximations

$$X_i^n \text{ of } u_f^n(x_i), \quad Y_j^n \text{ of } u_m^n(y_j), \quad Z_{i,j}^n \text{ of } c_{i,j}^n,$$

and

$$S_{f_i}^n, S_{m_j}^n \text{ of } s_{f_i}^n = s_f(x_i, t^n), s_{m_j}^n = s_m(y_j, t^n),$$

respectively. We approximate the density of couples with either partner of age h by zero, just for convenience in the proof of convergence. Since the time-step h will always be taken to be much smaller than the minimum age for reproduction or marriage, this is clearly a biological restriction too, just as much as it is at age 0.

Since the main part of our algorithm advances time—as well as age—two steps at once, we need to initialize the algorithm with the first two time-steps. We do this as follows: For $n = 0$ we use the initial age densities from the year Y_1 and let

$$\begin{cases} X_i^0 = u_{f_i}^0, \quad Y_j^0 = u_{m_j}^0, \quad 0 \leq i, j \leq L, \\ Z_{0,0}^0 = Z_{0,j}^0 = Z_{i,0}^0 = 0, \quad Z_{i,j}^0 = c_{i,j}^0, \quad 1 \leq i, j \leq L. \end{cases} \tag{4.7}$$

We use the composite trapezoidal rule to approximate the initial age densities of single females and males: For $0 \leq i, j \leq L$,

$$S_{f_i}^0 = X_i^0 - h\sum_{j=1}^{L-1} Z_{i,j}^0, \qquad S_{m_j}^0 = Y_j^0 - h\sum_{i=1}^{L-1} Z_{i,j}^0. \tag{4.8}$$

Even though the quadratures of the integrals are formally written as though they were (first order) composite endpoint rules, they are in fact equivalent to the (second order) composite trapezoidal rule because the terms corresponding to the summation index(es) equal to zero or L are always 0, by (4.7) and (4.2).

Next, for $n = 1$ we first define

$$Z^1_{0,j} = Z^1_{i,0} = Z^1_{1,j} = Z^1_{i,1} = 0, \qquad 0 \le i, j \le L+1, \tag{4.9}$$

and then use a linearized explicit Euler finite difference method of characteristics:

$$\begin{cases} \dfrac{Z^1_{i,j} - Z^0_{i-1,j-1}}{h} = -\sigma_{i-1,j-1} Z^0_{i-1,j-1} \\ \qquad\qquad + \mathcal{M}_{i-1,j-1}(S^0_{f_{i-1}}, S^0_{m_{j-1}}), \quad 2 \le i, j \le L+1, \\ \dfrac{X^1_i - X^0_{i-1}}{h} = -\mu_{f_{i-1}} X^0_{i-1}, \qquad 1 \le i \le L+1, \\ \dfrac{Y^1_j - Y^0_{j-1}}{h} = -\mu_{m_{j-1}} Y^0_{j-1}, \qquad 1 \le j \le L+1. \end{cases} \tag{4.10}$$

Finally, we use a composite trapezoidal rule to approximate births and age densities of single females and males:

$$\begin{cases} S^1_{f_i} = X^1_i - h \displaystyle\sum_{j=2}^{L} Z^1_{i,j}, \qquad 1 \le i \le L+1, \\ S^1_{m_j} = Y^1_j - h \displaystyle\sum_{i=2}^{L} Z^1_{i,j}, \qquad 1 \le j \le L+1, \\ X^1_0 = h^2 \displaystyle\sum_{i=2}^{L} \sum_{j=2}^{L} \beta_{f_{i,j}} Z^1_{i,j} + h \sum_{i=1}^{L} \beta^{(s)}_{f_i} S^1_{f_i}, \\ Y^1_0 = h^2 \displaystyle\sum_{i=2}^{L} \sum_{j=2}^{L} \beta_{m_{i,j}} Z^1_{i,j} + h \sum_{i=1}^{L} \beta^{(s)}_{m_i} S^1_{f_i}, \\ S^1_{f_0} = X^1_0, \qquad S^1_{m_0} = Y^1_0. \end{cases} \tag{4.11}$$

Here again, though the quadratures of the integrals are formally written as composite endpoint rules, they are in fact equivalent to composite trapezoidal rules because the terms corresponding to the summation index(es) equal to zero, one, or $L+1$ in $Z^1_{i,j}$ are always 0 in view of (4.9) and (4.2)–(4.4), as are those containing $\beta^{(s)}_{f_i}$, $\beta^{(s)}_{m_i}$, $\beta_{f_{i,j}}$, or $\beta_{m_{i,j}}$ with i or j equal to zero or $L+1$.

Then, in the recursive part of the algorithm, that is, for $2 \le n \le N$, we begin by setting the approximation of densities of couples with partners of age 0 or h equal to zero:

$$Z^n_{0,j} = Z^n_{i,0} = Z^n_{1,j} = Z^n_{i,1} = 0, \qquad 0 \le i, j \le L+n, \tag{4.12}$$

and we advance age and time for the densities of couples, females, and males using a Crank–Nicolson form of the finite difference method of characteristics:

$$\begin{cases} \dfrac{Z^n_{i,j} - Z^{n-2}_{i-2,j-2}}{2h} = -\sigma_{i-1,j-1}\dfrac{Z^n_{i,j} + Z^{n-2}_{i-2,j-2}}{2} \\ \qquad\qquad + \mathcal{M}_{i-1,j-1}(S^{n-1}_{f_{i-1}}, S^{n-1}_{m_{j-1}}), \quad 2 \le i, j \le L+n, \\ \dfrac{X^n_i - X^{n-2}_{i-2}}{2h} = -\mu_{f_{i-1}}\dfrac{X^n_i + X^{n-2}_{i-2}}{2}, \qquad 2 \le i \le L+n, \\ \dfrac{Y^n_j - Y^{n-2}_{j-2}}{2h} = -\mu_{m_{j-1}}\dfrac{Y^n_j + Y^{n-2}_{j-2}}{2}, \qquad 2 \le j \le L+n, \end{cases} \tag{4.13}$$

with Euler's method being employed for the first step of the age densities of females and males:

$$\begin{cases} \dfrac{X^n_1 - X^{n-1}_0}{h} = -\mu_{f_0} X^{n-1}_0, \\ \dfrac{Y^n_1 - Y^{n-1}_0}{h} = -\mu_{m_0} Y^{n-1}_0. \end{cases} \tag{4.14}$$

Finally, we use a composite trapezoidal rule to approximate births and age densities of single females and males:

$$\begin{cases} S^n_{f_i} = X^n_i - h \displaystyle\sum_{j=2}^{L+n-1} Z^n_{i,j}, \quad 1 \le i \le L+n, \\ S^n_{m_j} = Y^n_j - h \displaystyle\sum_{i=2}^{L+n-1} Z^n_{i,j}, \quad 1 \le j \le L+n, \\ X^n_0 = h^2 \displaystyle\sum_{i=2}^{L+n-1}\sum_{j=2}^{L+n-1} \beta_{f_{i,j}} Z^n_{i,j} + h \sum_{i=1}^{L+n-1} \beta^{(s)}_{f_i} S^n_{f_i}, \\ Y^n_0 = h^2 \displaystyle\sum_{i=2}^{L+n-1}\sum_{j=2}^{L+n-1} \beta_{m_{i,j}} Z^n_{i,j} + h \sum_{i=1}^{L+n-1} \beta^{(s)}_{m_i} S^n_{f_i}, \\ S^n_{f_0} = X^n_0, \qquad S^n_{m_0} = Y^n_0. \end{cases} \tag{4.15}$$

Just as before, the quadratures of the integrals—formally written as composite endpoint rules—are in fact equivalent to composite trapezoidal rules because the terms corresponding to the summation index(es) equal to zero, one, or $L+n$ in $Z^n_{i,j}$ are always 0 by (4.12) and (4.2)–(4.4), as are those containing $\beta^{(s)}_{f_i}$ or $\beta^{(s)}_{m_i}$ with i equal to zero or $L+n$.

4.2 Convergence of the method

We shall prove in this section that the algorithm described in the previous section leads to second order approximations when the coefficient functions are sufficiently smooth.

Let us introduce a convenient notation for the errors in the computed values of females, males, couples, and singles of each sex. For $g = f$ or m, $0 \leq i, j \leq L+n$, $0 \leq n \leq N$, let

$$\varepsilon_i^n = u_{f_i}^n - X_i^n, \quad \theta_j^n = u_{m_j}^n - Y_j^n, \quad \eta_{i,j}^n = c_{i,j}^n - Z_{i,j}^n, \quad \xi_{g_i}^n = s_{g_i}^n - S_{g_i}^n. \tag{4.16}$$

Theorem 4.1. *Let us assume that* $\mu_g \in C^0(0,\infty)$, $\beta_g \in C^2(0,\infty)$, $u_g \in C^3([0, M+T] \times [0,T])$ $(g = f, m)$, $\mathcal{M}$ *is uniformly Lipschitz continuous in its last two arguments* s_f *and* s_m, *with Lipschitz constant* K, $\sigma \in C^3([0, M+T] \times [0, M+T])$, *and* $c \in C^3([0, M+T] \times [0, M+T] \times [0,T])$. *Then, there exists a constant* $Q > 0$, *independent of* h, *such that, for* h *sufficiently small,*

$$\|\zeta\|_{l^\infty(l^1)}, \|\zeta\|_{l^\infty(l^\infty)} \leq Q h^2$$

for $\zeta = \varepsilon, \theta, \eta, \xi_f, \xi_m$, *where*

$$\|\zeta\|_{l^\infty(l^1)} = \max_{0 \leq n \leq N} \left\{ h \sum_{i=0}^{L+n} |\zeta_i^n| \right\}, \quad \|\zeta\|_{l^\infty(l^\infty)} = \max_{0 \leq n \leq N} \left\{ \max_{0 \leq i \leq L+n} \{|\zeta_i^n|\} \right\}$$

for $\zeta = \varepsilon, \theta, \xi_f, \xi_m$, *and*

$$\|\eta\|_{l^\infty(l^1)} = \max_{0 \leq n \leq N} \left\{ h^2 \sum_{i=0}^{L+n} \sum_{j=0}^{L+n} |\eta_{i,j}^n| \right\}, \quad \|\eta\|_{l^\infty(l^\infty)} = \max_{0 \leq n \leq N} \left\{ \max_{0 \leq i,j \leq L+n} \{|\eta_{i,j}^n|\} \right\}.$$

Proof. Note that using first order Taylor expansions we obtain, for $g = f$ or m and $1 \leq i \leq L+1$,

$$\frac{u_{g_i}^1 - u_{g_{i-1}}^0}{h} = -\mu_{g_{i-1}} u_{g_{i-1}}^0 + O\left(\left\| \frac{\partial^2 u_g}{\partial \nu^2} \right\|_{L^\infty} h \right), \tag{4.17}$$

while for $g = f$ or m, $2 \leq n \leq N$, and $2 \leq i \leq L+n$,

$$\frac{u_{g_i}^n - u_{g_{i-2}}^{n-2}}{2h} = -\mu_{g_{i-1}} \frac{u_{g_i}^n + u_{g_{i-2}}^{n-2}}{2} + O\left(\left[\left\| \frac{\partial^3 u_g}{\partial \nu^3} \right\|_{L^\infty} + \left\| \frac{\partial^2 u_g}{\partial \nu^2} \right\|_{L^\infty} \right] h^2 \right), \tag{4.18}$$

where $\nu = (\frac{1}{\sqrt{2}}, \frac{1}{\sqrt{2}})$ is the characteristic direction of the differential operator (4.1). Similarly, for $1 \leq i \leq L+1$,

$$\begin{aligned} \frac{c_{i,j}^1 - c_{i-1,j-1}^0}{h} &= -\sigma_{i-1,j-1} c_{i-1,j-1}^0 + \mathcal{M}_{i-1,j-1}\left(s_{f_{i-1}}^0, s_{m_{j-1}}^0\right) \\ &\quad + O\left(\left\| \frac{\partial^2 c}{\partial \nu^2} \right\|_{L^\infty} h \right), \end{aligned} \tag{4.19}$$

while for $2 \le n \le N$ and $2 \le i, j \le L+n$,

$$\frac{c^n_{i,j} - c^{n-2}_{i-2,j-2}}{2h} = -\sigma_{i-1} \frac{c^n_{i,j} + c^{n-2}_{i-2,j-2}}{2} + \mathcal{M}_{i-1,j-1}\left(s^{n-1}_{f_{i-1}}, s^{n-1}_{m_{j-1}}\right)$$
$$+ O\left(\left[\left\| \frac{\partial^3 c}{\partial \nu^3} \right\|_{L^\infty} + \left\| \frac{\partial^2 c}{\partial \nu^2} \right\|_{L^\infty}\right] h^2\right), \tag{4.20}$$

where $\nu = (\frac{1}{\sqrt{3}}, \frac{1}{\sqrt{3}}, \frac{1}{\sqrt{3}})$ is the characteristic direction of the differential operator in (2.8).

Also note that a first order Taylor expansion centered at $t = 0$ combined with the approximation properties of the composite trapezoidal rule imply that, for example, for $\beta_f^{(s)} \in C^2(0, \infty)$, $u_f \in C^2([0, M+nh) \times [0, M+nh))$ and compactly supported in that interval, we have

$$\int_0^\infty \beta_f^{(s)}(x)\, u_f(x, t^n)\, dx = h \sum_{i=1}^{L+n} \beta^{(s)}_{f_i}\, u^n_{f_i} + \frac{h}{2} \beta^{(s)}_{f_0}\, u^{n-1}_{f_0}$$
$$+ O\left(\left[\left\| \frac{\partial u_f(0,\cdot)}{\partial t} \right\|_{L^\infty} + \left\| \frac{\partial^2 \left[\beta_f^{(s)} u_f(\cdot, t^n)\right]}{\partial a^2} \right\|_{L^\infty}\right] h^2\right), \tag{4.21}$$

with similar expressions valid for the other integrals representing births and age densities of single females and males.

Now, combining (4.6), (4.7), (4.8), (4.12), and the expressions similar to (4.21) for the age densities of single females and males, we obtain the following bounds for the initial errors and for the boundary errors for the couples—corresponding to $n = 0$, $i = 0$ or 1, or $j = 0$ or 1:

$$\begin{cases} \varepsilon_i^0 = \theta_j^0 = \eta_{i,j}^0 = 0, \quad |\xi^0_{f_i}| = O(h^2), \quad |\xi^0_{m_j}| = O(h^2), \quad 0 \le i, j \le L, \\ \|\varepsilon^0\|_{l^1}, \|\theta^0\|_{l^1}, \|\eta^0\|_{l^1} = 0, \\ \eta^n_{0,j} = \eta^n_{i,0} = 0, \qquad 0 \le i, j \le L+n, \; 1 \le n \le N, \\ |\eta^n_{1,j}| = O(h^2), \quad |\eta^n_{i,1}| = O(h^2), \qquad 1 \le i, j \le L+n, \; 1 \le n \le N. \end{cases} \tag{4.22}$$

For $n = 1$ we combine (4.10) and (4.17) with (4.16) and (4.22) to derive the following error equations for the age densities of females and males for $1 \le i, j \le L+1$:

$$\begin{cases} \varepsilon_i^1 = (1 - \mu_{f_{i-1}} h)\varepsilon^0_{i-1} + O(h^2) = O(h^2), \\ \theta_j^1 = (1 - \mu_{m_{j-1}} h)\theta^0_{j-1} + O(h^2) = O(h^2). \end{cases} \tag{4.23}$$

Similarly, combining (4.9), (4.10), and (4.19) with (4.16) and (4.22), we see that for $2 \le i, j \le L+1$,

$$
\begin{aligned}
|\eta^1_{i,j}| &\le \big|\mathcal{M}_{i-1,j-1}(s^0_{f_{i-1}}, s^0_{m_{j-1}}) - \mathcal{M}_{i-1,j-1}(S^0_{f_{i-1}}, S^0_{m_{j-1}})\big|\, h \\
&\qquad + \left|(1-\sigma_{i-1,j-1}h)\,\eta^0_{i-1,j-1}\right| + O(h^2), \qquad (4.24)\\
&\le \left|\mathcal{M}_{i-1,j-1}(s^0_{f_{i-1}}, s^0_{m_{j-1}}) - \mathcal{M}_{i-1,j-1}(S^0_{f_{i-1}}, S^0_{m_{j-1}})\right| h + O(h^2).
\end{aligned}
$$

The combination of (4.22) and (4.24) leads, for $i, j \ge 1$, to the bound

$$|\eta^1_{i,j}| \le O(h^2), \tag{4.25}$$

while (4.23) readily yields the relations

$$|\varepsilon^1_i| \le O(h^2), \qquad |\theta^1_j| \le O(h^2), \quad 1 \le i, j \le L+1. \tag{4.26}$$

Next, note that (4.13), (4.16), and (4.20) give the following error equation for the age density of couples for $i, j, n \ge 2$:

$$
\begin{aligned}
\frac{\eta^n_{i,j} - \eta^{n-2}_{i-2,j-2}}{2h} &= -\sigma_{i-1,j-1}\frac{\eta^n_{i,j} + \eta^{n-2}_{i-2,j-2}}{2} + \mathcal{M}_{i-1,j-1}(s^{n-1}_{f_{i-1}}, s^{n-1}_{m_{j-1}}) \\
&\quad - \mathcal{M}_{i-1,j-1}(S^{n-1}_{f_{i-1}}, S^{n-1}_{m_{j-1}}) + O(h^2),
\end{aligned} \tag{4.27}
$$

while for the females and the males, in view of the comment that follows (4.15), we obtain from (4.13)–(4.18), using (4.21) and similar quadrature formulas for the other integrals in the expressions of the newborns,

$$
\left\{
\begin{aligned}
&\frac{\varepsilon^n_1 - \varepsilon^{n-1}_0}{h} = -\mu_{f_0}\varepsilon^{n-1}_0 + O(h),\\
&\frac{\varepsilon^n_i - \varepsilon^{n-2}_{i-2}}{2h} = -\mu_{f_{i-1}}\frac{\varepsilon^n_i + \varepsilon^{n-2}_{i-2}}{2} + O(h^2),\\
&\varepsilon^n_0 = h^2 \sum_{i=2}^{L+n-1}\sum_{j=2}^{L+n-1} \beta_{f_{i,j}}\,\eta^n_{i,j} + h\sum_{i=1}^{L+n-1} \beta^{(s)}_{f_i}\,\xi^n_{f_i} + O(h^2), \quad n \ge 0,\\
&\frac{\theta^n_1 - \theta^{n-1}_0}{h} = -\mu_{m_0}\theta^{n-1}_0 + O(h),\\
&\frac{\theta^n_j - \theta^{n-2}_{j-2}}{2h} = -\mu_{m_{j-1}}\frac{\theta^n_j + \theta^{n-2}_{j-2}}{2} + O(h^2),\\
&\theta^n_0 = h^2 \sum_{i=2}^{L+n-1}\sum_{j=2}^{L+n-1} \beta_{m_{i,j}}\,\eta^n_{i,j} + h\sum_{i=1}^{L+n-1} \beta^{(s)}_{m_i}\,\xi^n_{f_i} + O(h^2), \quad n \ge 0.
\end{aligned}
\right. \tag{4.28}
$$

Similarly, for the singles, (4.15), (4.16), and quadrature formulas similar to (4.21) lead to the relations

$$\begin{cases} \xi^n_{f_i} = \varepsilon^n_i - h \displaystyle\sum_{j=2}^{L+n-1} \eta^n_{i,j} + O(h^2), & 0 \le i \le L+n,\ 1 \le n \le N, \\ \xi^n_{m_j} = \theta^n_j - h \displaystyle\sum_{i=2}^{L+n-1} \eta^n_{i,j} + O(h^2), & 0 \le j \le L+n,\ 1 \le n \le N. \end{cases} \tag{4.29}$$

Setting

$$B = \max\left\{ \|\beta_f\|_{L^\infty}, \|\beta_m\|_{L^\infty}, \|\beta_f^{(s)}\|_{L^\infty}, \|\beta_m^{(s)}\|_{L^\infty} \right\},$$

we obtain from (4.28) the following relation for the errors in the newborns:

$$|\varepsilon^n_0|, |\theta^n_0| \le B\left(\|\eta^n\|_{l^1} + \|\xi^n_f\|_{l^1}\right) + O(h^2), \qquad 0 \le n \le N. \tag{4.30}$$

Also, it follows trivially from (4.29) that, for $i, j, n \ge 0$,

$$\begin{cases} |\xi^n_{f_i}| \le |\varepsilon^n_i| + \|\eta^n_{i,\cdot}\|_{l^1} + O(h^2), & \|\xi^n_f\|_{l^1} \le \|\varepsilon^n\|_{l^1} + \|\eta^n\|_{l^1} + O(h^2), \\ |\xi^n_{m_j}| \le |\theta^n_j| + \|\eta^n_{\cdot,j}\|_{l^1} + O(h^2), & \|\xi^n_m\|_{l^1} \le \|\theta^n\|_{l^1} + \|\eta^n\|_{l^1} + O(h^2), \end{cases} \tag{4.31}$$

where $\|\eta^n_{i,\cdot}\|_{l^1} = h\sum_{j=0}^{L+n} |\eta^n_{i,j}|$ and $\|\eta^n_{\cdot,j}\|_{l^1} = h\sum_{i=0}^{L+n} |\eta^n_{i,j}|$. Moreover, it follows from (4.22) and (4.25) that

$$\|\eta^1\|_{l^1} \le O(h^2), \tag{4.32}$$

which, combined with (4.30) and (4.31), yields

$$|\varepsilon^1_0|, |\theta^1_0| \le B\,\|\varepsilon^1\|_{l^1} + O\left(h^2\right). \tag{4.33}$$

This relation together with (4.26) gives, for $h < \frac{1}{2B}$,

$$\|\varepsilon^1\|_{l^1}, \|\theta^1\|_{l^1} \le O(h^2). \tag{4.34}$$

We rewrite (4.27) in the form

$$\begin{aligned} (1 + 2h\sigma_{i-1,j-1})\,\eta^n_{i,j} = (1 - 2h\sigma_{i-1,j-1})\,\eta^{n-2}_{i-2,j-2} + 2h\mathcal{M}_{i-1,j-1}(s^{n-1}_{f_{i-1}}, s^{n-1}_{m_{j-1}}) \\ - 2h\mathcal{M}_{i-1,j-1}(S^{n-1}_{f_{i-1}}, S^{n-1}_{m_{j-1}}) + O(h^3) \end{aligned}$$

and use the Lipschitz continuity of $\mathcal{M}$ to obtain, for $i, j, n \ge 2$ and $h < \frac{1}{2\|\sigma\|_{L^\infty}}$ (to ensure that $Z^n_{i,j} \ge 0$),

$$|\eta^n_{i,j}| \le |\eta^{n-2}_{i-2,j-2}| + 2Kh\left(|\xi^{n-1}_{f_{i-1}}| + |\xi^{n-1}_{m_{j-1}}|\right) + O(h^3). \tag{4.35}$$

Substituting (4.31) in (4.35) we arrive, for $i, j, n \geq 2$, at the estimate

$$|\eta^n_{i,j}| \leq |\eta^{n-2}_{i-2,j-2}| + Ch\left(|\varepsilon^{n-1}_{i-1}| + |\theta^{n-1}_{j-1}| + \|\eta^{n-1}_{i-1,\cdot}\|_{l^1} + \|\eta^{n-1}_{\cdot,j-1}\|_{l^1}\right) + O(h^3). \tag{4.36}$$

Multiplying (4.36) by h^2 and summing on i and j for $2 \leq i, j \leq L+n$, we see, using (4.22), that

$$\|\eta^n\|_{l^1} \leq \|\eta^{n-2}\|_{l^1} + Ch\left(\|\varepsilon^{n-1}\|_{l^1} + \|\theta^{n-1}\|_{l^1} + \|\eta^{n-1}\|_{l^1}\right) + O(h^3). \tag{4.37}$$

The value of the constant $C > 0$ need not be the same in each occurrence—it will just denote a generic positive constant independent of h. Proceeding along the same line on the second and fifth equations of (4.28) we arrive, for $i, j, n \geq 2$, at the relations

$$\begin{cases} |\varepsilon^n_i| \leq |\varepsilon^{n-2}_{i-2}| + O(h^3), \\ |\theta^n_j| \leq |\theta^{n-2}_{j-2}| + O(h^3), \end{cases} \tag{4.38}$$

while the first and fourth equations of (4.28) lead, for $n \geq 2$ and $h < \frac{1}{2\max\{\mu_{f_0}, \mu_{m_0}\}}$, to

$$\begin{cases} |\varepsilon^n_1| \leq |\varepsilon^{n-1}_0| + O(h^2), \\ |\theta^n_1| \leq |\theta^{n-1}_0| + O(h^2). \end{cases} \tag{4.39}$$

Multiplying (4.30), (4.38), and (4.39) by h and summing on i, j for $1 \leq i, j \leq L+n$, we see, using (4.31), that, for $n \geq 2$,

$$\begin{aligned} \|\varepsilon^n\|_{l^1} + \|\theta^n\|_{l^1} \leq \|\varepsilon^{n-2}\|_{l^1} + \|\theta^{n-2}\|_{l^1} + Ch\big(&\|\eta^n\|_{l^1} + \|\varepsilon^n\|_{l^1} \\ &+ \|\eta^{n-1}\|_{l^1} + \|\varepsilon^{n-1}\|_{l^1} + \|\theta^{n-1}\|_{l^1}\big) + O(h^3), \end{aligned}$$

which, added to (4.37), finally gives, for $n \geq 2$,

$$\begin{aligned} (1 - Ch)(\|\varepsilon^n\|_{l^1} + \|\theta^n\|_{l^1} + \|\eta^n\|_{l^1}) &\leq (\|\varepsilon^{n-2}\|_{l^1} + \|\theta^{n-2}\|_{l^1} + \|\eta^{n-2}\|_{l^1}) \\ &+ Ch(\|\varepsilon^{n-1}\|_{l^1} + \|\theta^{n-1}\|_{l^1} + \|\eta^{n-1}\|_{l^1}) + O(h^3). \end{aligned}$$

Adding the term $(1 - Ch)\left(\|\varepsilon^{n-1}\|_{l^1} + \|\theta^{n-1}\|_{l^1} + \|\eta^{n-1}\|_{l^1}\right)$ to both sides of this relation now leads to

$$\begin{aligned} &(1 - Ch)\left(\|\varepsilon^n\|_{l^1} + \|\theta^n\|_{l^1} + \|\eta^n\|_{l^1} + \|\varepsilon^{n-1}\|_{l^1} + \|\theta^{n-1}\|_{l^1} + \|\eta^{n-1}\|_{l^1}\right) \\ &\leq \left(\|\varepsilon^{n-1}\|_{l^1} + \|\theta^{n-1}\|_{l^1} + \|\eta^{n-1}\|_{l^1} + \|\varepsilon^{n-2}\|_{l^1} + \|\theta^{n-2}\|_{l^1} + \|\eta^{n-2}\|_{l^1}\right) + O(h^3), \end{aligned}$$

and recursive use of this formula into itself—together with (4.22), (4.32), and (4.34)—yields, for h sufficiently small, the relation

$$\|\varepsilon^n\|_{l^1}, \|\theta^n\|_{l^1}, \|\eta^n\|_{l^1} \leq O(h^2), \qquad n \geq 2, \tag{4.40}$$

whereby (4.31) results in

$$\left\|\xi_f^n\right\|_{l^1}, \left\|\xi_m^n\right\|_{l^1} \leq O(h^2), \qquad n \geq 2. \tag{4.41}$$

Combining (4.30), (4.39), (4.40), and (4.41), we see that, for $n \geq 2$ and $i, j = 0$ or 1,

$$|\varepsilon_i^n|, |\theta_j^n| \leq O(h^2). \tag{4.42}$$

In view of (4.42), when we use (4.38) recursively in itself together with (4.22), (4.26), and (4.33), we derive the thesis for $\zeta = \varepsilon$ and θ. Consequently, we can rewrite (4.36) in the form

$$|\eta_{i,j}^n| \leq |\eta_{i-2,j-2}^{n-2}| + Ch\left(\left\|\eta_{i-1,\cdot}^{n-1}\right\|_{l^1} + \left\|\eta_{\cdot,j-1}^{n-1}\right\|_{l^1}\right) + O(h^3), \quad i, j, n \geq 2. \tag{4.43}$$

Multiplying this relation by h and summing on j, $2 \leq j \leq L+n$, we see, in view of (4.32) and (4.40), that

$$\|\eta_{i,\cdot}^n\|_{l^1} \leq \|\eta_{i-2,\cdot}^{n-2}\|_{l^1} + Ch\|\eta_{i-1,\cdot}^{n-1}\|_{l^1} + O(h^3),$$

and adding $\|\eta_{i-1,\cdot}^{n-1}\|_{l^1}$ to both sides of this relation we see that

$$\begin{aligned}\|\eta_{i,\cdot}^n\|_{l^1} + \|\eta_{i-1,\cdot}^{n-1}\|_{l^1} &\leq \|\eta_{i-2,\cdot}^{n-2}\|_{l^1} + (1+Ch)\|\eta_{i-1,\cdot}^{n-1}\|_{l^1} + O(h^3) \\ &\leq (1+Ch)\left(\|\eta_{i-2,\cdot}^{n-2}\|_{l^1} + \|\eta_{i-1,\cdot}^{n-1}\|_{l^1}\right) + O(h^3).\end{aligned}$$

Iterating this formula in itself we are led, using (4.22) and (4.32), to the bound

$$\left\|\eta_{i,\cdot}^n\right\|_{l^1} \leq O(h^2), \qquad i, n \geq 1.$$

Analogously, we see that

$$\left\|\eta_{\cdot,j}^n\right\|_{l^1} \leq O(h^2), \qquad j, n \geq 1.$$

The combination of these two relations with (4.43) yields the estimate

$$|\eta_{i,j}^n| \leq |\eta_{i-2,j-2}^{n-2}| + O(h^3), \qquad i, j, n \geq 2,$$

which, when iterated in itself, leads to the thesis for $\zeta = \eta$, in view of (4.22) and (4.25). Finally, for $\zeta = \xi_g$, $g = f, m$, the result follows trivially from (4.31). □

4.3 Using U.S. Census data and Vital Statistics Reports to estimate model parameters

In order to use the models we presented in Chapter 2 to make projections about the population of the United States, several types of data are needed for the simulator (a computer program that makes projections by numerically solving the integrodifferential systems in the models). First, we need the *initial data*—that is, the age distributions of females, males, and couples in the year Y_1 that will be used as a basis for the projections. For females and males this is the easiest data to obtain, since it is readily available from the U.S. Bureau of the Census. For the year 2000, we can even find these data online at

http://factfinder.census.gov/servlet/DTTable?_bm=y&-geo_id=01000US&-ds_name=DEC_2000_SF1_U&-_lang=en&-mt_name=DEC_2000_SF1_U_PCT012&-_sse=on

We present these values in Table 4.1.

For each sex the age data are grouped in one hundred and three cohorts: one hundred one-year cohorts of individuals of age in the interval $[A, A+1)$ for integers $0 \leq A \leq 99$, two five-year cohorts of individuals of age in the interval $[A, A+5)$ for $A = 100$ and $A = 105$, and one residual cohort of individuals of age at least 110 years. Figure 4.1 shows these age distributions in graphic form.

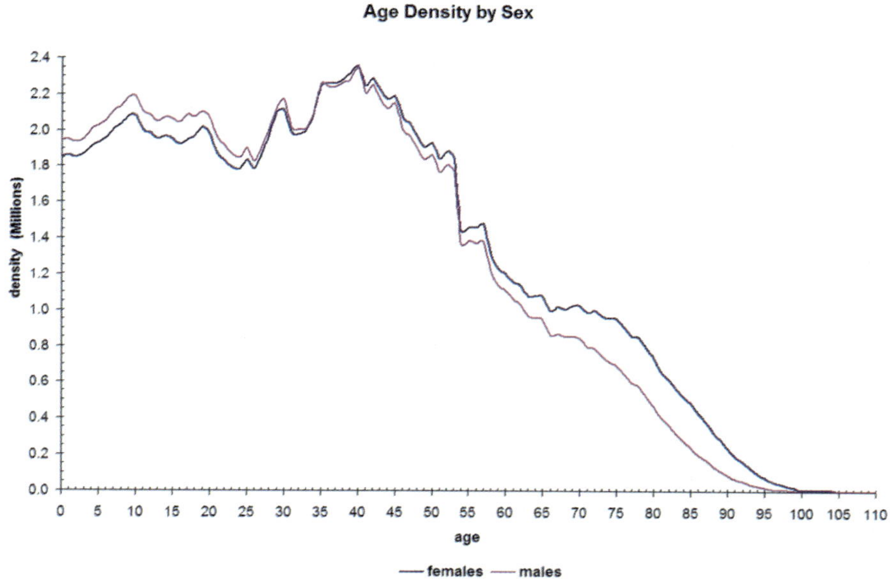

Figure 4.1. *Age density of U.S. females and males* (2000).

Unfortunately, the data for the age distribution of couples in the U.S. are no longer available in the detail required by the model. We quote verbatim from the *Federal Register*, December 15, 1995 (Volume 60, Number 241, pp. 64437–64438):

> "Summary: Beginning January 1, 1996, the availability of marriage and divorce data collected by the National Center for Health Statistics (NCHS), CDC, will change. NCHS will continue to collect marital status in all of its population surveys, will continue to obtain detailed information on out-of-wedlock births, and will work with States to obtain summary counts of marriages and divorces. However, detailed data from States participating in the marriage and divorce components of the Vital Statistics Cooperative Program (VSCP) will no longer be obtained. This change is being made to prioritize programs in a period of tightened resource constraints."

Table 4.1. *Age density of the U.S. population by sex in the year* 2000.

Age	# Females	# Males	Age	# Females	# Males
< 1 year	1,856,631	1,949,017	52 years	1,891,651	1,815,785
1 year	1,867,477	1,953,105	53 years	1,856,617	1,778,423
2 years	1,851,456	1,938,990	54 years	1,445,145	1,372,415
3 years	1,873,836	1,958,993	55 years	1,463,741	1,386,859
4 years	1,915,665	2,010,658	56 years	1,462,265	1,375,187
5 years	1,934,031	2,031,072	57 years	1,479,824	1,384,196
6 years	1,961,488	2,058,217	58 years	1,317,443	1,222,709
7 years	2,008,279	2,109,868	59 years	1,237,235	1,139,778
8 years	2,041,401	2,137,829	60 years	1,208,384	1,111,560
9 years	2,081,029	2,186,291	61 years	1,159,548	1,061,679
10 years	2,082,812	2,191,244	62 years	1,137,207	1,033,865
11 years	2,006,936	2,108,157	63 years	1,081,948	971,203
12 years	1,988,614	2,087,228	64 years	1,081,733	958,320
13 years	1,956,842	2,054,008	65 years	1,079,260	950,651
14 years	1,972,671	2,079,560	66 years	996,164	864,156
15 years	1,954,277	2,065,127	67 years	1,022,372	874,079
16 years	1,926,439	2,048,582	68 years	1,008,370	856,145
17 years	1,954,732	2,091,280	69 years	1,027,017	855,331
18 years	1,972,745	2,078,853	70 years	1,030,658	844,517
19 years	2,020,693	2,107,162	71 years	989,752	798,517
20 years	1,978,228	2,071,220	72 years	1,000,532	791,164
21 years	1,875,409	1,965,673	73 years	973,735	751,433
22 years	1,837,099	1,921,549	74 years	959,852	717,281
23 years	1,798,182	1,875,400	75 years	955,776	695,865
24 years	1,787,269	1,853,972	76 years	908,794	647,773
25 years	1,838,640	1,905,899	77 years	861,039	599,742
26 years	1,787,277	1,832,383	78 years	852,548	579,368
27 years	1,874,853	1,914,947	79 years	793,200	521,708
28 years	1,974,005	2,010,807	80 years	740,352	467,013
29 years	2,107,801	2,134,724	81 years	665,502	406,546
30 years	2,115,732	2,174,238	82 years	616,747	364,815
31 years	1,991,793	2,019,782	83 years	565,774	317,289
32 years	1,985,244	2,008,877	84 years	522,095	279,234
33 years	2,008,556	2,018,017	85 years	485,320	244,874
34 years	2,087,294	2,100,855	86 years	430,173	204,981
35 years	2,250,497	2,265,621	87 years	383,810	173,520
36 years	2,263,639	2,247,529	88 years	326,086	139,395
37 years	2,266,938	2,250,122	89 years	287,928	113,731
38 years	2,285,731	2,268,083	90 years	238,226	89,678
39 years	2,321,163	2,287,341	91 years	197,406	68,980
40 years	2,358,828	2,352,606	92 years	163,780	54,437
41 years	2,253,642	2,213,034	93 years	129,373	39,693
42 years	2,290,677	2,256,543	94 years	101,421	29,537
43 years	2,229,419	2,178,451	95 years	76,998	21,097
44 years	2,180,195	2,128,468	96 years	57,976	14,704
45 years	2,190,345	2,151,115	97 years	42,536	10,308
46 years	2,077,993	2,009,570	98 years	29,199	6,804
47 years	2,043,564	1,976,128	99 years	21,960	5,202
48 years	1,975,473	1,909,672	100–104 years	36,717	8828
49 years	1,915,523	1,843,021	105–109 years	2,836	685
50 years	1,936,877	1,871,638	≥ 110 years	844	544
51 years	1,847,534	1,769,463	**TOTAL**	143,368,343	138,053,563

The 1990 data for the age distribution of couples, even though they precede this reduction in data availability, seem not to exist. For earlier U.S. census years, such as 1970 and 1980, the age densities of married couples needed in the model are available in U.S. Bureau of the Census Reports [118, 121]. We show the density of married couples in 1980 by age of each spouse as a three-dimensional surface in Figure 4.2 and as a contour plot in Figure 4.3. Since the available data are grouped in five-year age-cohorts of each spouse, the numbers used for the figures are those from the data divided by five. The last age-cohorts for each spouse contain all individuals 80 years of age or older.

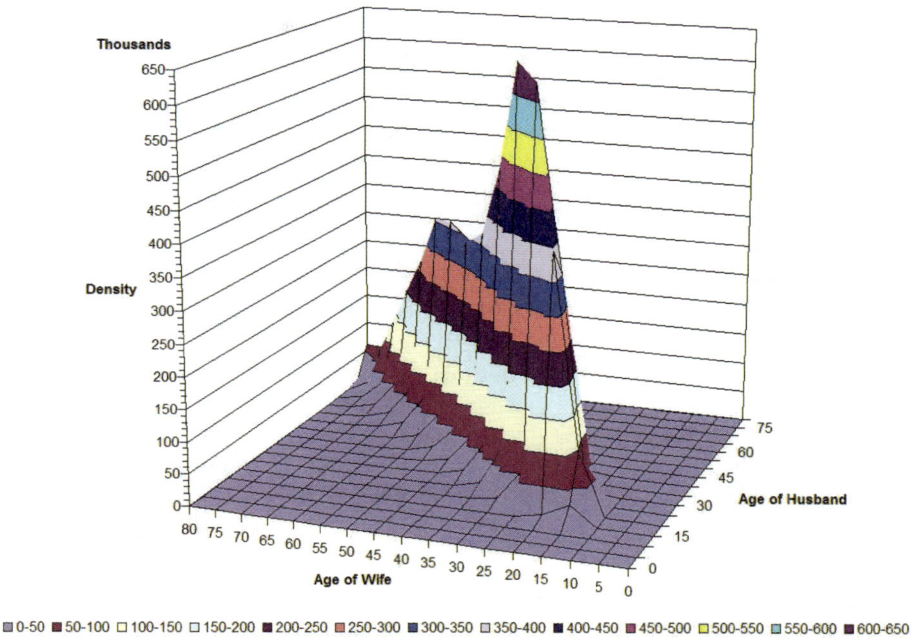

Figure 4.2. *Age density of U.S. married couples* (1980).

Also, we need some vital rates that depend on one age variable only, namely the mortality rates for females and males and the birth rates for girls and boys from single mothers. The former are easily available [97]—the latter not quite. In fact what is easily available is the total birth rate by age of the mother (irrespective of marital status or sex of the newborn) as well as the total birth rate by age of the mother for single women [96].

Mortality rates for each sex come grouped and averaged in cohorts of five or more years, giving piecewise constant functions (*step functions*) rather than the smoother functions our convergence theorem requires, as shown next in Figures 4.4 and 4.5 for the year 2000.

The corresponding probability of survival functions, π_f and π_m, are smoother thanks to the integration that defines them, as seen in Figure 4.6.

We show in Figure 4.7 the female and male probabilities of survival to a given age up to 50 years of age, a detail of Figure 4.6.

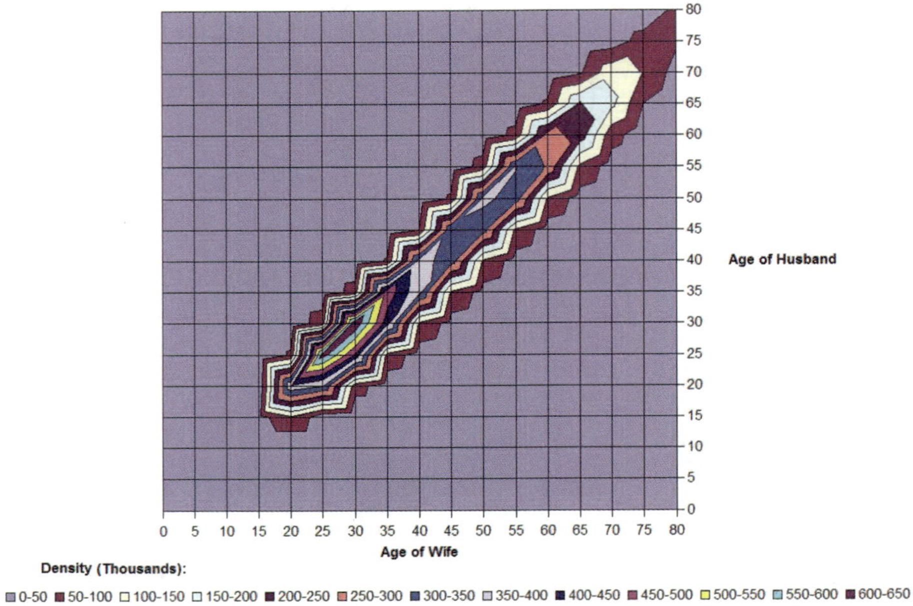

Figure 4.3. *Contour plot of the age density of U.S. married couples* (1980).

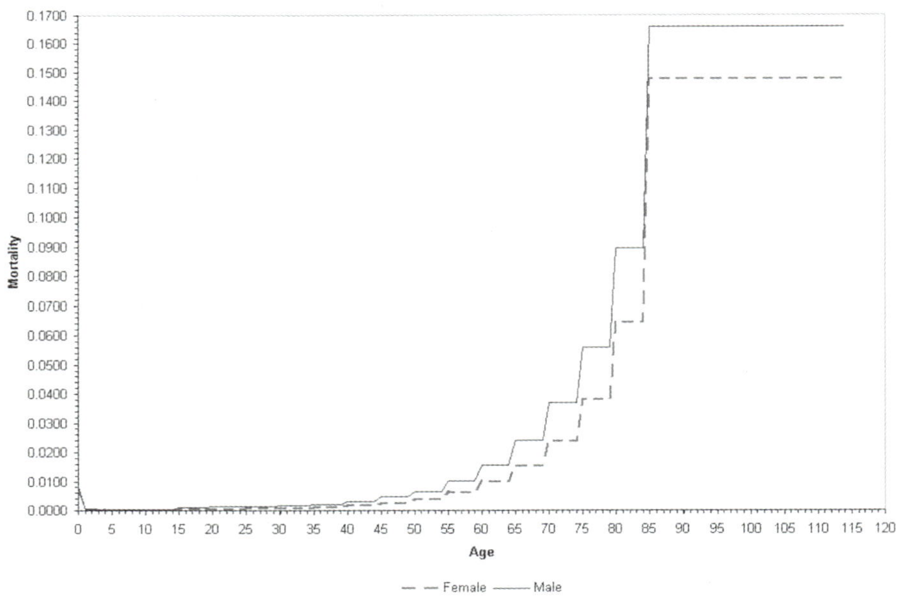

Figure 4.4. *Mortality rates by sex in the U.S.* (2000).

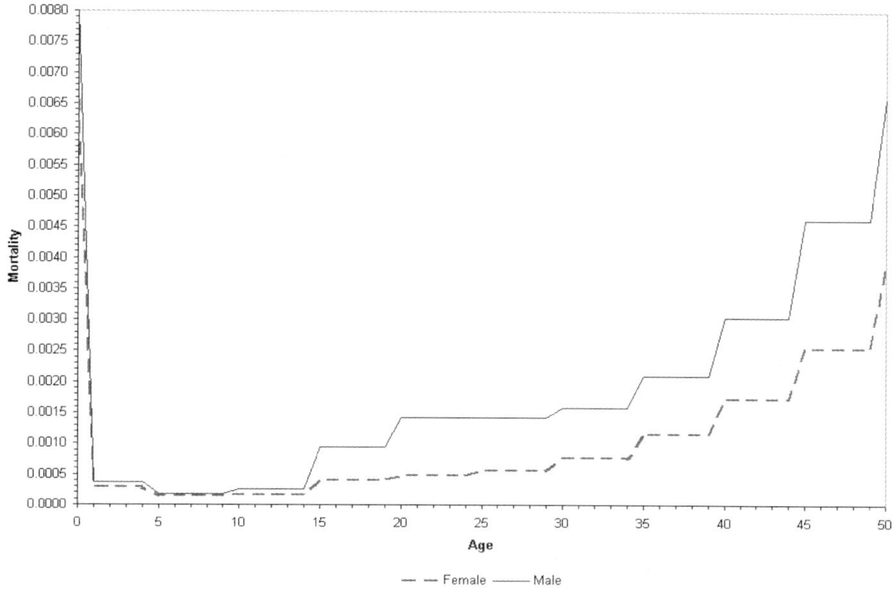

Figure 4.5. *Mortality rates in the U.S. to* 50 *years of age* (2000).

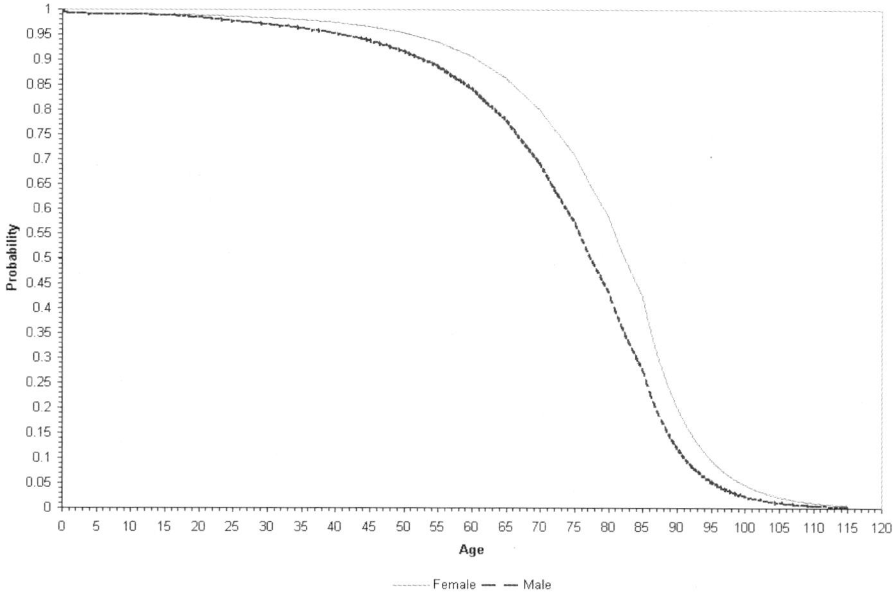

Figure 4.6. *Probability of survival at birth by sex* (2000).

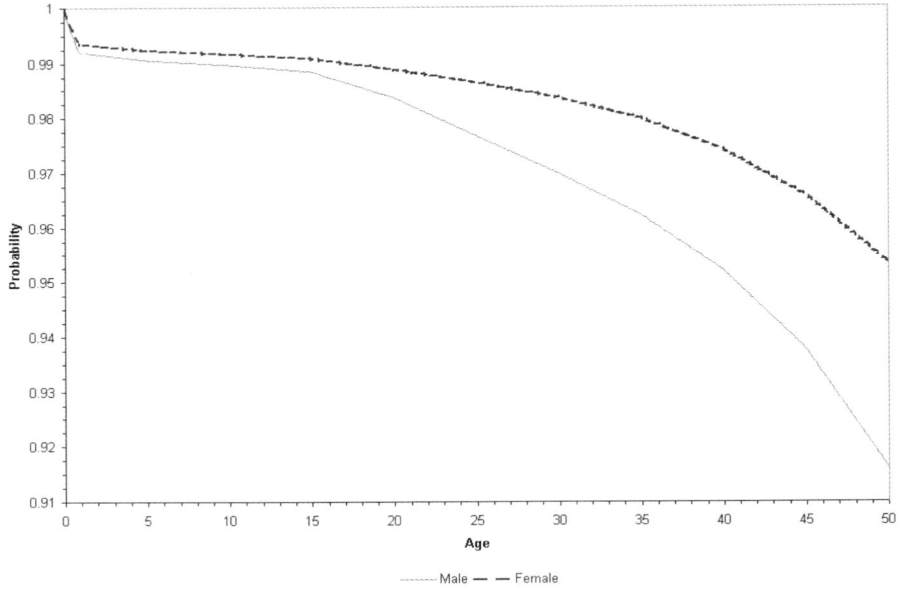

Figure 4.7. *Probability of survival by sex to* 50 *years of age* (2000).

It is apparent in Figure 4.7 that the probability of survival after age 15 drops much more sharply for males than for females. Despite the larger birth rate for boys than for girls, this leads to larger female age-cohorts than the corresponding ones for males for all cohorts past age 35, as seen in Figure 4.1.

Concerning birth rates from single mothers by age of mother and sex of newborn, at present National Vital Statistics Reports [96] provide total birth rates, $\beta_i^{(s)}$, for unmarried women by age of the mother but, unfortunately, not separated by sex of the newborn. This means that no data are directly available to generate the functions $\beta_f^{(s)}(x)$ and $\beta_m^{(s)}(x)$. In order to separate the newborns by sex we can use the sex ratio at birth uniformly across all age-cohorts of mothers and fathers, as well as marital status of the mother. For example, in the year 2000 there were 1,981,845 girls and 2,076,969 boys born in the U.S., giving a sex ratio at birth of

$$\gamma = \frac{2{,}076{,}969}{1{,}981{,}845} \approx 1.0479977. \tag{4.44}$$

Similarly, in the year 1990 there were 2,028,717 girls and 2,129,495 boys born in the U.S., giving a sex ratio at birth of

$$\gamma = \frac{2{,}129{,}495}{2{,}028{,}717} \approx 1.0496757.$$

This should not introduce much error in the simulation, since the sex ratio at birth has varied by less than 1% in the period 1940–2000, with a high value of 1.055 in 1940, and there is no reason to believe it will be different in the future. Using this ratio we compute the birth rates of girls (respectively, boys) from unmarried women in each age-cohort as follows:

$$\beta_{f_i}^{(s)} = \frac{1}{1+\gamma}\,\beta_i^{(s)} \quad \left(\text{respectively, } \beta_{m_i}^{(s)} = \frac{\gamma}{1+\gamma}\,\beta_i^{(s)}\right).$$

Finally, we need several demographic rates that depend on two age variables, none of which are directly available. These are the birth rates of girls and boys from married couples, $\beta_f(x, y)$ and $\beta_m(x, y)$, the divorce/annulment rate by age of each spouse, $\delta(x, y)$, and the intrinsic age-specific marriage rate $\rho(x, y)$. Concerning the birth rates, we can find in the National Vital Statistics Reports [96] the total number of newborn girls and the total number of newborn boys by age of each parent (respectively, G_{ij} and B_{ij}), as well as the total birth rate for married women for each age-cohort of the mother, β_i. From these we compute the birth rates of girls and boys from married couples as follows.

We first compute the sizes of the married-women age-cohorts, C_{f_i}, by summing the sizes of cohorts of couples (C_{ij}) with wife in the corresponding age-cohort over all age-cohorts of husbands,

$$C_{f_i} = \textstyle\sum_j C_{ij}, \tag{4.45}$$

and we compute the number of offspring of each sex born from mothers in each age-cohort (respectively, G_i and B_i) by summing the numbers G_{ij} (respectively, B_{ij}) over all age-cohorts of fathers:

$$G_i = \textstyle\sum_j G_{ij} \text{ (respectively, } B_i = \textstyle\sum_j B_{ij}). \tag{4.46}$$

We then compute the birth rates of girls and boys from married couples for each age-cohort of mother and father,

$$\beta_{f_{ij}} = \frac{G_{ij}\,\beta_i\,C_{f_i}}{C_{ij}\,G_i}, \qquad \beta_{m_{ij}} = \frac{B_{ij}\,\beta_i\,C_{f_i}}{C_{ij}\,B_i}, \tag{4.47}$$

which is equivalent to

$$\beta_{f_{ij}} C_{ij} = \frac{G_{ij}\beta_i C_{f_i}}{G_i}, \qquad \beta_{m_{ij}}\,C_{ij} = \frac{B_{ij}\,\beta_i\,C_{f_i}}{B_i},$$

expressing the total number of newborn girls (respectively, boys) from married couples in each age-cohort of mother and father as the product of the total number of newborn girls (respectively, boys) from parents (married or not) in that same age-cohort of mother and father, multiplied by the ratio of married women to all women in the corresponding age-cohort. In this way we ensure that the total number of girls (respectively, boys) born to married women in each age-cohort is the correct one, i.e., the product of the total number of girls (respectively, boys) born to all women in that age-cohort and the ratio of married women to all women in that cohort.

Concerning the divorce/annulment rate and the intrinsic age-specific marriage rate, we can find in the National Vital Statistics Reports [95] the numbers of divorces/annulments, A_{f_i} and A_{m_j}, as well as the age-specific marriage rates, ρ_{f_i} and ρ_{m_j}, for each age-cohort of wife or husband, and the total number of annulments/divorces, A. The latter is important because there is a significant fraction of all annulments in which the age of one or both spouses is not reported. We use these data to define $\delta(x, y)$ and $\rho(x, y)$ using a product structure:

$$\delta_{ij} = \frac{A_{f_i}}{C_{f_i}} \frac{A_{m_j}}{C_{m_j}} \frac{A}{C}, \qquad \rho_{ij} = \rho_{f_i} \rho_{m_j},$$

where $C_{m_j} = \sum_i C_{ij}$ and $C = \sum_j C_{m_j} = \sum_i C_{f_i}$ are, respectively, the sizes of the married-men age-cohorts and the total number of married couples, computed analogously to (4.45).

Concerning the birth rates from single mothers by age of mother and sex of newborn, we present an alternative way for computing these without using the sex ratio at birth (4.44). The rates for girls are computed for each age-cohort of single mothers as the product of the total birth rate for that cohort and the ratio of newborn girls to all newborns to all women in that age-cohort:

$$\beta^{(s)}_{f_i} = \beta^{(s)}_{i} \frac{G_i}{G_i + B_i}.$$

Then, the rates for boys for each age-cohort of the mother are computed as the difference between the total rates and the rates for girls:

$$\beta^{(s)}_{m_i} = \beta^{(s)}_{i} - \beta^{(s)}_{f_i}.$$

4.4 Estimation of parameters from insufficient and/or inconsistent data

Data collected in population censuses and local and State Bureaus of Health Statistics, though plentiful, are not always available in the form the models require. The available data concerning births and marriages are sometimes inconsistent and cannot be used in the form given in the tables without some processing.

In 1–3 below, we discuss now some of the problems with the data and the approach we took to solve them.

1. *Births to married women of ages* 10–14. In the United States the number of births to women under the age of 15 can usually be obtained for any year from natality tables in the U.S. Vital Statistics [120]. We note that there is a difference between the total number of births to women in this age group and the number of births to unmarried women of ages 10–14. This suggests that some of the mothers in this age group are married. However, when we check Volume II of the U.S. Vital Statistics [121], which reports statistics about marriage and divorce, we find a statement that marriages in this age bracket are so scarce that they can safely be taken as zero. As a result, we have a positive number of births to married women of age under 15 but we don't have any married women of this age. To resolve this paradox, or inconsistency, we make the following assumption:

number of married women aged 10–14 = number of births in this age group
− number of births to unmarried women in this age group.

This assumption is reasonable since most marriages at this young age take place because of a pregnancy. Moreover, the vast majority of mothers in this age group have only one child.

2. *The vital rates in the statistics.* Besides the *number* of births, the statistics for the United States also provide vital *rates*, including birth rates. It is unclear what data are used for computing these rates since some of the reported numbers lead to inconsistencies. For example, for 1980 the cohort of women of ages 25–29 years is reported to consist of 2,111,744 never-married women, 51,751 widowed women, and 890,566 divorced women. This produces a total of 3,054,061 single women, the sum of these three figures (see [121, Table 1]). The 1980 U.S. Census of the population (Table 1-34 in [120]) reports 99,583 births to unmarried women. From the ratio of these two values we find the birth rate to unmarried women in this cohort to be 32.6 per 1000. However, the reported birth rate from the same source (Table 1-32 in [120]) is 34.0—a difference of 5%. In order to avoid an inconsistency, we computed the birth rates as ratios using the given data. It should be noted that birth rates are especially prone to this variation since different sources give different numbers for the married/unmarried women.

3. *Undistributed data in the tables.* Some of the Vital Statistics Tables have a column labeled "not stated" for the number of those who did not report their status. This problem is particularly frequent for those densities and rates that are cross tabulated by two, three, or more characteristics, such as the number of newborns tabulated by sex, and by age of the father and of the mother. The numbers in these columns are often in the range of 10%–15% of the total. If this column is simply neglected, the initial data used for a projection can be severely underestimated (by at least 10%).

Consequently, for the birth rates of boys and girls by age of father, we adjusted the values given in Table 1-55 of [120] by distributing the births corresponding to the column of fathers who did not report their age among all age-cohorts of the fathers. We distributed the numbers from the "not stated" column proportionally to the number of births for these groups of fathers using the available birth rates by age of father—from which we computed the modified total number of births in every age group of the father. For example, for the cohort of fathers aged 30–34, we used the male birth rate from fathers in this cohort, β^{30}, together with the total number of newborn boys, B, and the number of boys born to fathers who did not report their ages, $B_{\text{n.r.}}$—both reported in the Vital Statistics Tables—to compute a corrected male birth rate from fathers aged 30–34 as follows:

$$\beta(y) = \beta^{30}\left(1 + \frac{B_{\text{n.r.}}}{B}\right), \quad 30 \le y < 35.$$

4.5 Simulation of the population of the U.S. using data from two-sex life tables

We present in Tables 4.2 and 4.3 the results for ten-year simulations of the population of the U.S., respectively, from 1970 to 1980 [1] and from 1980 to 1990. The marriage function used was the true harmonic mean without age preferences, i.e., $g(x, y) = h(x, y) = \rho(x, y)$ in (2.14). The time dependence of the vital rates is modeled by linear interpolation of their values in 1970 and 1980 (respectively, in 1980 and 1990).

Also, to illustrate the importance of the time dependence of the vital rates, we present in Table 4.4 the age densities of females and males obtained from those in 1970 in a ten-year simulation using constant-in-time vital rates at their 1970 values.

Table 4.2. *Relative errors by sex for* 1980 *(linear-in-time vital rates).*

	MALES			FEMALES		
Age bracket (yrs.)	**Calculated** (1000s)	**Actual** (1000s)	**Error** (%)	**Calculated** (1000s)	**Actual** (1000s)	**Error** (%)
0–4	8482	8362	−1.4	8062	7986	−0.9
5–9	8931	8539	−4.6	8509	8161	−4.3
10–14	8659	9316	7.1	8345	8926	6.5
15–19	10122	10755	5.9	9787	10413	6.0
20–24	10467	10663	1.8	9997	10655	6.2
25–29	9538	9705	1.7	9153	9816	6.8
30–34	7596	8677	12.5	8180	8884	7.9
35–39	6418	6862	6.5	6748	7104	5.0
40–44	5425	5708	5.0	5756	5961	3.4
45–49	5194	5388	3.6	5565	5702	2.4
50–54	5419	5621	3.6	5916	6089	2.9
55–59	5215	5482	4.9	5887	6133	4.0
60–64	4493	4670	3.8	5255	5418	3.0
65–69	3668	3903	6.0	4577	4880	6.2
70–74	2749	2854	3.7	3773	3945	4.3
75–79	1766	1848	4.4	2809	2946	4.7
80–84	1029	1019	−1.0	1890	1916	1.4
85 and over	755	682	−10.8	1604	1559	−2.9
Total	105928	110053	3.8	111811	116493	4.0

Table 4.3. *Relative errors by sex in* 1990 *(linear-in-time vital rates).*

	MALES			FEMALES		
Age bracket (yrs.)	**Calculated** (1000s)	**Actual** (1000s)	**Error** (%)	**Calculated** (1000s)	**Actual** (1000s)	**Error** (%)
0–4		9392			8962	
5–9		8539			8837	
10–14	8318	8767	5.1	7955	8347	4.7
15–19	8486	9103	6.8	8137	8651	5.9
20–24	9199	9676	4.9	8886	9345	4.9
25–29	10572	10696	1.2	10351	10617	2.5
30–34	10461	10877	3.8	10581	10986	3.7
35–39	9488	9902	4.2	9726	10061	3.3
40–44	8440	8692	2.9	8770	8924	1.7
45–49	6603	6811	3.0	6961	7062	1.4
50–54	5389	5515	2.3	5775	5836	1.0
55–59	4922	5034	2.2	5421	5497	1.4
60–64	4897	4947	1.0	5644	5669	0.4
65–69	4462	4532	1.5	5467	5579	2.0
70–74	3450	3409	−1.2	4569	4586	0.4
75–79	2534	2400	−5.6	3765	3722	−1.2
80–84	1542	1366	−12.8	2654	2568	−3.4
85 and over	960	858	−12.0	2310	2222	−4.0
Total	117037	121239	3.5	123566	127470	3.1

Table 4.4. *Relative errors by sex in* 1980 *(constant-in-time vital rates).*

Age bracket		MALES			FEMALES	
	Calculated	Actual	Error	Calculated	Actual	Error
(yrs.)	(1000s)	(1000s)	(%)	(1000s)	(1000s)	(%)
0–4	10222	8362	−22.2	9743	7986	−22.0
5–9	9381	8539	−9.9	8946	8161	−9.6
10–14	8653	9316	7.1	8341	8926	6.5
15–19	10115	10755	6.0	9567	10413	7.3
20–24	10459	10663	1.9	9713	10655	8.8
25–29	9529	9705	1.8	8984	9816	8.5
30–34	7589	8677	12.5	8149	8884	8.3
35–39	6399	6862	6.7	6733	7104	5.2
40–44	5397	5708	5.0	5737	5961	3.8
45–49	5157	5388	4.3	5541	5702	2.8
50–54	5369	5621	4.5	5886	6089	3.3
55–59	5135	5482	6.3	5849	6133	4.6
60–64	4394	4670	5.9	5213	5418	3.8
65–69	3561	3903	8.8	4514	4880	7.5
70–74	2648	2854	7.2	3696	3945	6.3
75–79	1684	1848	8.8	2703	2946	8.2
80–84	972	1019	4.7	1782	1916	7.0
85 and over	758	680	−11.4	1529	1559	1.9
Total	107423	110053	2.4	112715	116493	3.2

When the distribution of marriages by age of wife and husband is known for the years Y_1 and Y_2, we can compute in addition the preference distributions g and h that go into the marriage function (2.14). For example, g_{ij}, the preference of females in the ith age-cohort for males in the jth age-cohort for the year Y_1, is found as the ratio of the number of marriages in the corresponding age-cohorts of wife and husband in that year, M_{ij}, to the number of single females in the ith age-cohort, $S_{f_i} = X_i - C_{f_i}$:

$$g_{ij} = \frac{M_{ij}}{S_{f_i}}.$$

We show in Figures 4.8 and 4.9 the graph of the female preference distribution for the year 1970, respectively, as a three-dimensional surface and a contour plot.

The way these functions are calculated ensures that the total number of marriages in each age-cohort of wife and husband for the corresponding year is represented exactly. To see this let us represent by h_{ij} the preference of males in the jth age-cohort for females in the ith age-cohort, by C_{m_j} the number of married males in the jth age-cohort, and by S_{m_j} the number of single males in the jth age-cohort for the year Y_1. Then, $C_{m_j} = \sum_i C_{ij}$, $S_{m_j} = Y_j - C_{m_j}$, and $h_{ij} = \frac{M_{ij}}{S_{m_j}}$. Then, we have

$$\mathcal{M}_{ij}(S_{f_i}, S_{m_j}) = 2\,\frac{g_i j S_{f_i} h_{ij} S_{m_j}}{g_i j S_{f_i} + h_{ij} S_{m_j}}$$

$$= 2\,\frac{M_{ij} M_{ij}}{M_{ij} + M_{ij}} = M_{ij}.$$

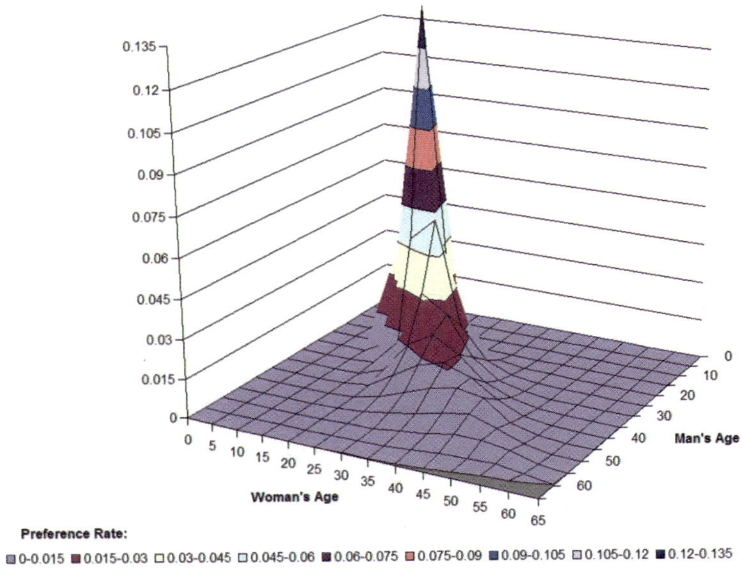

Figure 4.8. *Female preference distribution* (1970).

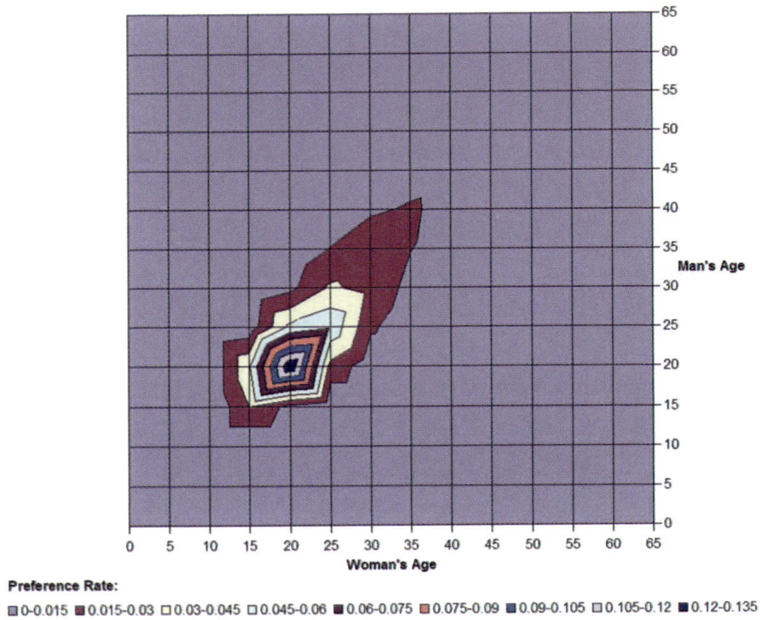

Figure 4.9. *Contour plot of female preference distribution* (1970).

Table 4.5. *Relative errors by sex for* 1980 *(using mating preferences).*

	MALES			FEMALES		
Age bracket (yrs.)	**Calculated** (1000s)	**Actual** (1000s)	**Error** (%)	**Calculated** (1000s)	**Actual** (1000s)	**Error** (%)
0–4	8365	8362	−0.04	7948	7986	−0.5
5–9	8756	8539	−2.5	8337	8161	−2.2
10–14	8659	9316	7.1	8345	8926	6.5
15–19	10122	10755	5.9	9787	10413	6.0
20–24	10467	10663	1.8	9997	10655	6.2
25–29	9538	9705	1.7	9153	9816	6.8
30–34	7596	8677	12.5	8180	8884	7.9
35–39	6418	6862	6.5	6748	7104	5.0
40–44	5425	5708	5.0	5756	5961	3.4
45–49	5194	5388	3.6	5565	5702	2.4
50–54	5419	5621	3.6	5916	6089	2.9
55–59	5215	5482	4.9	5887	6133	4.0
60–64	4493	4670	3.8	5255	5418	3.0
65–69	3668	3903	6.0	4577	4880	6.2
70–74	2749	2854	3.7	3773	3945	4.3
75–79	1766	1848	4.4	2809	2946	4.7
80–84	1029	1019	−1.0	1890	1916	1.4
85 and over	755	682	−10.8	1604	1559	−2.9
Total	105636	110053	4.0	111526	116493	4.3

Next we present in Table 4.5 the results of a ten-year simulation of the population of the U.S. from 1970 to 1980, using these age-preference distributions.

There is no difference between the errors obtained when using age preferences and when not using them (Tables 4.2 and 4.5) for age-cohorts older than 10 years of age. This is due to the fact that age preferences only affect marriages, and, as a consequence, couples distributions and births. In fact, comparing the first two age-cohorts in Tables 4.2 and 4.5 for either sex, it is striking to see the relative errors reduced by 44%–96% when using age preferences.

We show in Figures 4.10 and 4.11 the evolution of the age density of females resulting from the ten-year simulation just described for the harmonic mean with age preferences.

In order to illustrate the limitation of the model for long-term projections, we present finally in Table 4.6 the age densities of females and males obtained from those in 1970 in a twenty-year simulation, using the true harmonic mean with preferences marriage function and linear-in-time vital rates that interpolate the 1970 and 1980 values.

4.6 A simulation-based approach to comparing marriage functions

We performed an extensive comparison of ten-year projections of the population of the U.S. using the numbers given by the 1970 census as initial data and letting the model with births outside marriage described in section 2.4 make projections of the age distributions of females and males—as well as of couples—for 10 years later using many different marriage functions in the family of generalized means described in Chapter 2.

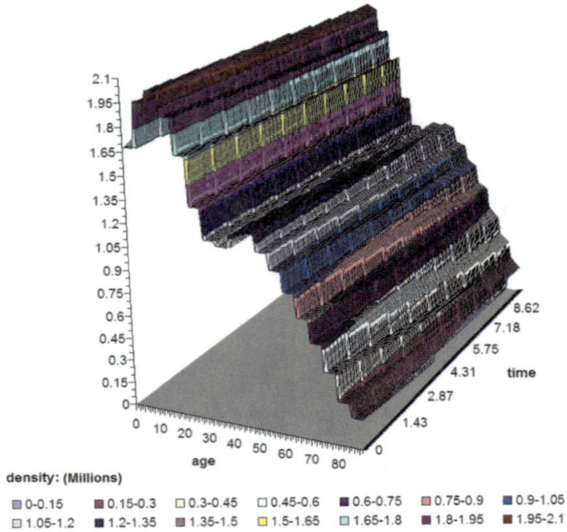

Figure 4.10. *Female age density* (1970–1980).

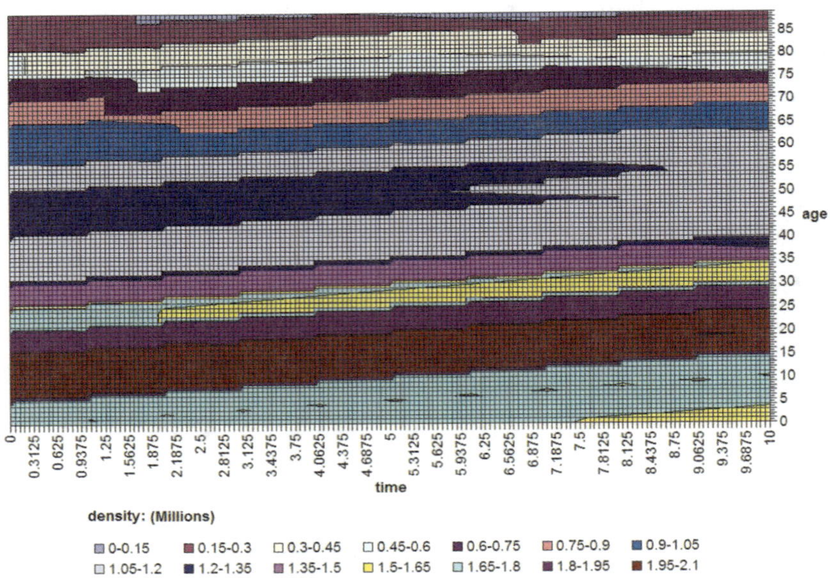

Figure 4.11. *Contour plot of female age density* (1970–1980).

Table 4.6. *Relative errors by sex for* 1990 *(time-extrapolated vital rates).*

Age bracket (yrs.)	MALES Calculated (1000s)	MALES Actual (1000s)	MALES Error (%)	FEMALES Calculated (1000s)	FEMALES Actual (1000s)	FEMALES Error (%)
0–4	5867	9392	37.5	5572	8962	37.8
5–9	7381	8539	13.6	7013	8837	20.6
10–14	8323	8767	5.1	7919	8347	5.1
15–19	8701	9103	4.4	8321	8651	3.8
20–24	8547	9676	11.7	8334	9345	10.8
25–29	9952	10696	7.0	9764	10617	8.0
30–34	10280	10877	5.5	9948	10986	9.4
35–39	9356	9902	5.5	9085	10061	9.7
40–44	7432	8692	14.5	8096	8924	9.3
45–49	6203	6811	8.9	6628	7062	6.1
50–54	5131	5515	7.0	5587	5836	4.3
55–59	4759	5034	5.5	5304	5497	3.5
60–64	4744	4947	4.1	5502	5669	2.9
65–69	4234	4532	6.6	5275	5579	5.4
70–74	3279	3409	3.8	4462	4586	2.7
75–79	2276	2400	5.1	3538	3722	4.9
80–84	1364	1366	0.1	2526	2568	1.6
85 and over	865	858	−0.8	2201	2222	0.9
Total	108732	121239	10.3	115221	127470	9.6

In order to compare the performance of the various marriage functions in the chosen family of generalized means with preferences, we only need to compare their projections for couples and newborns of each sex, since changing only the marriage function does not alter the projections of the aging process in the model for either sex. Therefore, to compare different marriage functions we look at the relative error in their projected total numbers of newborn girls, newborn boys, and couples. This relative error is found by subtracting the results of our simulations from the numbers taken from the actual distributions (given by the 1980 U.S. Census [119, 120]) and dividing the differences by these numbers.

We ran our simulations with a time-step $h = 0.0625$ years (≈ 22.8 days) to obtain the predicted distribution of males, females, and couples in 1980, for several choices of the marriage function. We then compared the performance of different marriage functions obtained with various values for the parameters α and β on the basis of how they project births (for boys and girls separately) and how they project the total number of couples.

First, we simulated the limiting case $\alpha \to -\infty$, which, as indicated in Chapter 2, corresponds to the marriage function (E5)

$$\mathcal{M}(x, y; s_f, s_m) = \min\{g(x, y)s_f, h(x, y)s_m\}.$$

For this marriage function the relative errors in the numbers of newborn girls, newborn boys, and couples are, respectively,

$$5.2\%, \quad 4.8\%, \quad \text{and} \quad 6.1\%.$$

For $\beta = 0$ we obtain—independently of the value of α—the female dominance function

$$\mathcal{M}(x, y; s_f, s_m) = g(x, y)s_f.$$

The errors in this case for the numbers of newborn girls, newborn boys, and couples are, respectively,

$$1.8\%, \quad 1.4\%, \quad \text{and} \quad 3.4\%.$$

Similarly, for $\beta = 1$ we obtain—independently of the value of α—the male dominance function

$$\mathcal{M}(x, y; s_f, s_m) = h(x, y)s_m.$$

The errors in this case for the numbers of newborn girls, newborn boys, and couples are, respectively,

$$0.3\%, \quad 0.2\%, \quad \text{and} \quad 2.7\%.$$

For other values of α and β we present, in Tables 4.7–4.10, the results from some of the simulations we performed. We indicate the errors in newborn females, males, and total number of couples for several values of α and in each table we vary β from 0.2 to 0.8 in one-tenth increments.

Looking at the errors from all these simulations we see that, as α increases, the errors become smaller and, in fact, they are monotonically decreasing for any fixed value of β. Analogously, for a fixed value of α, we see that in all the tables the errors decrease as β increases. We realize that the minimum function is worst among all marriage functions we tested, with relative errors in the numbers of newborn females, newborn males, and total couples between two and twenty-four times as large as those for almost all other marriage functions we used.

We also see that the male dominance function ($\beta = 1$) gives consistently the smallest errors of all marriage functions tested. However, this marriage function does not satisfy several of the general conditions imposed on the marriage functions based on theoretical or demographic reasons. In particular, it does not satisfy the heterosexuality condition of not allowing marriages when no singles of one sex are present, and it does not satisfy the consistency condition of not allowing more marriages than the total number of singles available from either sex.

The observed behavior in the errors is probably due to reasons external to the model and does not allow us to draw immediate conclusions about the goodness of fit for the various functions. In fact the best performing function in our simulations can hardly be perceived as the "best marriage function" since it fails to satisfy some of the basic properties of marriage functions, as well as some intuitive expectations about the sociology of marriage.

In conclusion we would like to make the following observations.

1. We readily see that the errors in the total numbers of births projected by the worst performing function—the minimum function—are one order of magnitude larger than those projected by the best performing function—the male dominance—and of the same magnitude as the average error. Hence, this test compares favorably to most of the other tests based on the quality of the fit where the error differences between the various functions are much smaller than the average error. The fact that there are noticeable differences in the errors leads us to the conclusion that, in order to make projections with the two-sex model, we must choose the marriage function that provides the best fit.

Table 4.7. *Relative errors in numbers of newborns and couples.*

$$\mathcal{M}(x, y; s_f, s_m) = \frac{g(x, y)s_f\, h(x, y)s_m}{\sqrt{\beta\left[g(x, y)s_f\right]^2 + (1-\beta)\left[h(x, y)s_m\right]^2}} \qquad (\alpha = -2)$$

β	**0.2**	**0.3**	**0.4**	**0.5**	**0.6**	**0.7**	**0.8**
Girls	2.5%	2.5%	2.5%	2.4%	2.1%	1.9%	1.6%
Boys	2.0%	2.1%	2.0%	1.9%	1.7%	1.5%	1.1%
Couples	4.1%	4.2%	4.2%	4.2%	4.1%	4.0%	3.8%

Table 4.8. *Relative errors in numbers of newborns and couples.*

$$\mathcal{M}(x, y; s_f, s_m) = \frac{g(x, y)s_f\, h(x, y)s_m}{\beta g(x, y)s_f + (1-\beta)h(x, y)s_m} \qquad (\alpha = -1)$$

(weighted harmonic mean)

β	**0.2**	**0.3**	**0.4**	**0.5**	**0.6**	**0.7**	**0.8**
Girls	2.1%	2.1%	2.0%	1.9%	1.8%	1.5%	1.2%
Boys	1.7%	1.7%	1.6%	1.5%	1.3%	1.1%	0.8%
Couples	3.8%	3.9%	3.9%	3.8%	3.8%	3.7%	3.5%

Table 4.9. *Relative errors in numbers of newborns and couples.*

$$\mathcal{M}(x, y; s_f, s_m) = \frac{g(x, y)s_f\, h(x, y)s_m}{\left[\beta\sqrt{g(x, y)s_f} + (1-\beta)\sqrt{h(x, y)s_m}\right]^2} \qquad (\alpha = -\tfrac{1}{2})$$

β	**0.2**	**0.3**	**0.4**	**0.5**	**0.6**	**0.7**	**0.8**
Girls	2.0%	1.9%	1.8%	1.7%	1.5%	1.3%	1.0%
Boys	1.5%	1.5%	1.4%	1.3%	1.0%	0.9%	0.6%
Couples	3.7%	3.7%	3.7%	3.7%	3.6%	3.5%	3.3%

Table 4.10. *Relative errors in numbers of newborns and couples.*

$$\mathcal{M}(x, y; s_f, s_m) = \left[h(x, y)s_m\right]^{\beta}\left[g(x, y)s_f\right]^{1-\beta} \qquad (\alpha \to 0^-)$$

(weighted geometric mean)

β	**0.2**	**0.3**	**0.4**	**0.5**	**0.6**	**0.7**	**0.8**
Girls	1.8%	1.7%	1.6%	1.5%	1.3%	1.1%	0.9%
Boys	1.4%	1.3%	1.2%	1.0%	0.9%	0.7%	0.4%
Couples	3.5%	3.5%	3.5%	3.4%	3.3%	3.2%	3.1%

2. We can also suggest that if the source of the biggest contributor to the error in the age-cohorts beyond 15 years of age is removed—that is, if immigration could be adequately modeled—then the resulting model could be used to help support or question theoretical deliberations about the form of the marriage function.

3. An observation worth making is that the introduction of preferences leads to a significant improvement in the ability of the marriage functions to make projections of the total number of couples. Of course this observation has its logical explanation: preferences introduce more information from data into the marriage function and hence improve the fit. This is also the reason why Henry's and McFarland's models perform better than traditional marriage functions.

4. Another question that certainly deserves attention is the following: The male dominance function is certainly not the best theoretical marriage function but is it not still possible that the fact that it leads to better projections is a consequence of an actual demographic phenomenon? Indeed, if the marriage market obeys market laws, the outcome—the number of marriages—should be more heavily influenced by the scarcer sex. And the fact is that after the age of 20—which includes most marriagable ages—females outnumber males. This observation somewhat contradicts the general perception that females have greater impact on marriages and that a marriage function (if not symmetric) should be slightly female dominant [69]. The latter perception is based on the assumption that women have more motivation to get married, while the former is a logical consequence of the larger availability of partners.

5. The observed pattern in the behavior of the errors can be used as a criterion for choosing an appropriate marriage function to improve the projective properties of the simulator. In fact, as can be seen from our experiments, the total number of newborn girls and boys and the total number of couples can be obtained with very good accuracy, provided that an appropriate marriage function is used. In fact a new, more complex marriage function can be proposed by assigning for the number of marriages in each bride–groom age category the projection from the marriage function that results in the largest number of marriages.

In conclusion we would like to stress that an abundant literature exists discussing the problem of the form of the marriage function. Various formulas and procedures have been proposed for the computation of the number of marriages from the numbers of single male and female individuals. But the problem of choosing the "right" marriage function in general—or in each particular situation—still remains open and, quite possibly, will never be settled.

Chapter 5
Age Profiles and Exponential Growth

Il burattino, ritornato in cittá, cominció a contare i minuti a uno a uno e, quando gli parve che fosse l'ora, riprese subito la strada che menava al campo dei miracoli. E mentre camminava con passo frettoloso, il cuore gli batteva forte e gli faceva tic tac, tic tac, come un orologio da sala quando corre davvero. E intanto pensava dentro di se: "E se invece di mille monete ne trovassi sui rami dell'albero duemila? E se invece di duemila ne trovassi cinquemila? E se invece di cinquemila ne trovassi centomila? Oh, che bel signore allora che diventerei!"[1]

—Carlo Collodi, *Le avventure di Pinocchio,* 1867

The sequence of golden coins in Pinocchio's dream sounds much like it is growing exponentially. Exponential growth is the main aspect of our model because the homogeneity of the problem—due to the properties of the marriage function—forces the model to support exponential growth of the population. This is observed in the two-sex model without age structure. In this chapter we shall describe the existing results on a two-sex analogue of the stable population theory that is available for the single population model introduced in Chapter 1. In contrast to the one-sex case, where the McKendrick–von Foerster model is linear, two-sex models are nonlinear because it is necessary to include coupling.

The role of steady states in homogeneous systems is played by the persistent solutions. Establishing the existence of these is the first step toward a global behavior result. A significant result in that direction was obtained by Prüss and Schappacher in [107]. They show the existence of persistent solutions with a positive two-sex growth rate for

[1] Once he was back in town, the puppet began to count minutes and, when he thought it was time, he took his way to the field of miracles. And, while walking in a hurry, his heart was doing tick tock, tick tock, like a clock in a true hurry. Meanwhile, he was thinking: "What if, instead of one thousand coins, I found on the tree two thousand of them? And what if, instead of two thousand, I found five thousand of them? And if instead of five thousand I found one hundred thousand? Oh, what a nice rich man I would become!"

Hadeler's two-sex age-structured model with a specific marriage function, namely function (E9). The same result is established with a general homogeneous marriage function for the Fredrickson–Hoppensteadt model with constant sex ratio at birth in [82]. Recently, Prüss and Schappacher's result for Hadeler's model was also shown to be valid with a general homogeneous marriage function by Zacher [129]. In the next section we present the first result on the existence of persistent solutions with a two-sex growth rate that may be positive or negative. For consistency we establish this in the space of continuous functions, though all previously mentioned results do it in the space of integrable functions. We consider the Fredrickson–Hoppensteadt model with a constant sex ratio at birth.

There are two major ways that a homogeneous system can be treated—directly or after normalization with a functional. In this chapter we have chosen to work with the second approach. Actually, we shall pursue a description focused on the ratio of the age distribution of the females, males, couples, and singles to the total population size—that is, we shall consider as variables the age profiles of the females, males, couples, and singles. Thus, we are led to consider and analyze steady profiles rather than steady populations. We obtain existence of persistent solutions that show separation between the age distribution and the population growth.

5.1 The age profiles equations

We consider the Fredrickson–Hoppensteadt model of Chapter 2 in the autonomous case. That is, throughout this chapter we assume that there is no explicit dependence of the rates and of the marriage function upon the time variable. In particular, the basic function $\mathcal{F}$ defining the marriage function is independent of t. Moreover, we assume that the marriage function satisfies property (P3), i.e., it is homogeneous of degree one. In fact all the following considerations depend on this property.

We shall consider the total population

$$P(t) = \int_0^\omega u_f(x,t)dx + \int_0^\omega u_m(y,t)dy,$$

the female, male, and couples age profiles, respectively,

$$\mathcal{U}_f(x,t) = \frac{u_f(x,t)}{P(t)}, \quad \mathcal{U}_m(y,t) = \frac{u_m(y,t)}{P(t)}, \quad \mathcal{C}(x,y,t) = \frac{C(x,y,t)}{P(t)},$$

and the total population growth rate

$$\lambda(t) = \frac{P'(t)}{P(t)} = \int_0^\omega \int_0^\omega (\beta_f(x,y) + \beta_m(x,y))\mathcal{C}(x,y,t)dxdy$$
$$- \int_0^\omega \mu_f(x)\mathcal{U}_f(x,t)dx - \int_0^\omega \mu_m(y)\mathcal{U}_m(y,t)dy.$$

Following a calculation similar to that performed in the single population case (see (1.11) and (1.12)) we arrive at the following problem for the variables $\mathcal{U}_f, \mathcal{U}_m, \mathcal{C}, \lambda$:

$$\begin{cases} \dfrac{\partial \mathcal{U}_f}{\partial t} + \dfrac{\partial \mathcal{U}_f}{\partial x} = -\lambda(t)\mathcal{U}_f(x,t) - \mu_f(x)\mathcal{U}_f(x,t), \\ \mathcal{U}_f(0,t) = \displaystyle\int_0^\omega \int_0^\omega \beta_f(x,y)\mathcal{C}(x,y,t)dxdy, \end{cases} \tag{5.1}$$

$$\begin{cases} \dfrac{\partial \mathcal{U}_m}{\partial t} + \dfrac{\partial \mathcal{U}_m}{\partial y} = -\lambda(t)\mathcal{U}_m(y,t) - \mu_m(y)\mathcal{U}_m(y,t), \\ \mathcal{U}_m(0,t) = \displaystyle\int_0^\omega \int_0^\omega \beta_m(x,y)\mathcal{C}(x,y,t)dxdy, \end{cases} \tag{5.2}$$

$$\begin{cases} \dfrac{\partial \mathcal{C}}{\partial t} + \dfrac{\partial \mathcal{C}}{\partial x} + \dfrac{\partial \mathcal{C}}{\partial y} = -\lambda(t)\mathcal{C}(x,y,t) - \sigma(x,y)\mathcal{C}(x,y,t) + \mathcal{M}(x,y,t;\mathcal{S}_f,\mathcal{S}_m), \\ \mathcal{C}(x,0,t) = \mathcal{C}(0,y,t) = 0, \end{cases} \tag{5.3}$$

$$\int_0^\omega \mathcal{U}_f(x,t)dx + \int_0^\omega \mathcal{U}_m(y,t)dy = 1, \tag{5.4}$$

where $\mathcal{S}_f$ and $\mathcal{S}_m$ are the age profiles of the single females and males, respectively:

$$\begin{aligned} \mathcal{S}_f(x,t) &= \mathcal{U}_f(x,t) - \int_0^\omega \mathcal{C}(x,y,t)dy, \\ \mathcal{S}_m(y,t) &= \mathcal{U}_m(y,t) - \int_0^\omega \mathcal{C}(x,y,t)dx. \end{aligned} \tag{5.5}$$

The main goal of this formulation is to separate the equations for the age profiles from that for the total population. Actually, once we have a solution for (5.1)–(5.4) the evolution of the total population is given by

$$P(t) = P(0)e^{\int_0^t \lambda(s)ds}. \tag{5.6}$$

The system (5.1)–(5.5) is nonlinear and not homogeneous. Therefore, it has stationary solutions that correspond to persistent solutions of the original system. We consider next the stationary version of (5.1)–(5.4), which we study under the additional assumption of a fixed sex ratio at birth:

$$\beta_f(x,y) = (1-\gamma)\beta(x,y), \quad \beta_m(x,y) = \gamma\beta(x,y), \tag{5.7}$$

where $\gamma \in (0,1)$ is the proportion of males among the newborns. This is a fairly natural assumption and it leads to a significant simplification in the treatment of the problem. Under this assumption, the search for stationary profiles is reduced to the study of the coupled system

$$\begin{cases} \mathcal{U}_f'(x) = -\lambda \mathcal{U}_f(x) - \mu_f(x)\mathcal{U}_f(x), \\ \mathcal{U}_f(0) = (1-\gamma)\int_0^\omega \int_0^\omega \beta(x,y)\mathcal{C}(x,y)dxdy, \end{cases} \tag{5.8}$$

$$\begin{cases} \mathcal{U}_m'(y) = -\lambda \mathcal{U}_m(y) - \mu_m(y)\mathcal{U}_m(y), \\ \mathcal{U}_m(0) = \gamma \int_0^\omega \int_0^\omega \beta(x,y)\mathcal{C}(x,y)dxdy, \end{cases} \tag{5.9}$$

$$\begin{cases} \dfrac{\partial \mathcal{C}}{\partial x} + \dfrac{\partial \mathcal{C}}{\partial y} = -\lambda \mathcal{C}(x,y) - \sigma(x,y)\mathcal{C}(x,y) + \mathcal{M}(x,y;\mathcal{S}_f,\mathcal{S}_m), \\ \mathcal{C}(x,0) = \mathcal{C}(0,y) = 0, \end{cases} \tag{5.10}$$

$$\int_0^\omega \mathcal{U}_f(x)dx + \int_0^\omega \mathcal{U}_m(y)dy = 1, \tag{5.11}$$

where

$$\mathcal{S}_f(x) = \mathcal{U}_f(x) - \int_0^\omega \mathcal{C}(x,y)dy, \quad \mathcal{S}_m(y) = \mathcal{U}_m(y) - \int_0^\omega \mathcal{C}(x,y)dx.$$

We note that a stationary state $\mathcal{U}_f^*, \mathcal{U}_m^*, \mathcal{C}^*, \lambda^*$ for the profiles corresponds to pure Malthusian growth for the total population (see (5.6)),

$$P^*(t) = P^*(0)e^{\lambda^* t},$$

and to the following "persistent solution" of the original Fredrickson–Hoppensteadt dynamic problem (2.6)–(2.8):

$$\begin{aligned} u_f^*(x,t) &= e^{\lambda^* t}\mathcal{U}_f^*(x), \\ u_m^*(y,t) &= e^{\lambda^* t}\mathcal{U}_m^*(y), \\ c^*(x,y,t) &= e^{\lambda^* t}\mathcal{C}^*(x,y). \end{aligned}$$

In order to study the previous problem, we need to perform further transformations, which we shall describe in the next section.

5.2 Preliminary transformation of the problem

To simplify the search for solutions of (5.8)–(5.11) we further transform this problem to prepare it for the proof of existence. For this purpose we first set

$$B = \int_0^\omega \int_0^\omega \beta(x,y)\mathcal{C}(x,y)dxdy \tag{5.12}$$

and then take advantage of the homogeneity of the problem to rescale the variables into new ones:

$$\bar{\mathcal{U}}_f(x) = \frac{1}{B}\mathcal{U}_f(x), \quad \bar{\mathcal{U}}_m(y) = \frac{1}{B}\mathcal{U}_m(y), \quad \bar{\mathcal{C}}(x, y) = \frac{1}{B}\mathcal{C}(x, y)\,. \tag{5.13}$$

Next, we consider the new equations they must satisfy:

$$\begin{cases} \bar{\mathcal{U}}_f'(x) = -\lambda\bar{\mathcal{U}}_f(x) - \mu_f(x)\bar{\mathcal{U}}_f(x), \\ \bar{\mathcal{U}}_f(0) = 1 - \gamma, \end{cases} \tag{5.14}$$

$$\begin{cases} \bar{\mathcal{U}}_m'(y) = -\lambda\bar{\mathcal{U}}_m(y) - \mu_m(y)\bar{\mathcal{U}}_m(y), \\ \bar{\mathcal{U}}_m(0) = \gamma, \end{cases} \tag{5.15}$$

$$\begin{cases} \dfrac{\partial\bar{\mathcal{C}}}{\partial x} + \dfrac{\partial\bar{\mathcal{C}}}{\partial y} = -\lambda\bar{\mathcal{C}}(x, y) - \sigma(x, y)\bar{\mathcal{C}}(x, y) + \mathcal{M}(x, y; \bar{\mathcal{S}}_f, \bar{\mathcal{S}}_m), \\ \bar{\mathcal{C}}(x, 0) = \bar{\mathcal{C}}(0, y) = 0, \end{cases} \tag{5.16}$$

$$\int_0^\omega \int_0^\omega \beta(x, y)\bar{\mathcal{C}}(x, y)dxdy = 1, \tag{5.17}$$

where

$$\bar{\mathcal{S}}_f(x) = \bar{\mathcal{U}}_f(x) - \int_0^\omega \bar{\mathcal{C}}(x, y)dy, \quad \bar{\mathcal{S}}_m(y) = \bar{\mathcal{U}}_m(y) - \int_0^\omega \bar{\mathcal{C}}(x, y)dx.$$

As a matter of fact, (5.14)–(5.17) are equivalent to (5.8)–(5.11) because, once we find a solution $\bar{\mathcal{U}}_f, \bar{\mathcal{U}}_m, \bar{\mathcal{C}}$ of the former, we have a solution of the latter by setting (see (5.11) and (5.13))

$$B = \frac{1}{\int_0^\omega \bar{\mathcal{U}}_f(x)dx + \int_0^\omega \bar{\mathcal{U}}_m(y)dy}$$

and

$$\mathcal{U}_f(x) = B\,\bar{\mathcal{U}}_f(x), \quad \mathcal{U}_m(y) = B\,\bar{\mathcal{U}}_m(y), \quad \mathcal{C}(x, y) = B\,\bar{\mathcal{C}}(x, y).$$

Now we see that system (5.14)–(5.15) is decoupled from (5.16)–(5.17) so that we can solve it explicitly and further reduce the problem. More specifically, we have (see Chapter 3 for the meaning of $\pi_f(x)$, $\pi_m(y)$, and $\pi_c(x, y; z)$)

$$\bar{\mathcal{U}}_f(x) = (1 - \gamma)e^{-\lambda x}\pi_f(x), \quad \bar{\mathcal{U}}_m(y) = \gamma e^{-\lambda y}\pi_m(y), \tag{5.18}$$

so that

$$\begin{aligned} \bar{\mathcal{S}}_f(x) &= (1 - \gamma)e^{-\lambda x}\pi_f(x) - \int_0^\omega \bar{\mathcal{C}}(x, y)dy, \\ \bar{\mathcal{S}}_m(y) &= \gamma e^{-\lambda y}\pi_m(y) - \int_0^\omega \bar{\mathcal{C}}(x, y)dx, \end{aligned} \tag{5.19}$$

and, in view of (5.16)–(5.17), we are only left with the following two equations in the variables $\lambda, \bar{C}$:

$$\bar{C}(x, y) = \int_0^{x \wedge y} e^{-\lambda\tau} \pi_c(x, y; \tau) \mathcal{M}(x - \tau, y - \tau; \bar{S}_f, \bar{S}_m) d\tau, \tag{5.20}$$

$$\int_0^{\omega} \int_0^{\omega} \beta(x, y) \bar{C}(x, y) dx dy = 1, \tag{5.21}$$

where we have used (5.19). These equations should be enough to treat our original problem, but, as a matter of fact, it is more convenient to switch to a new formulation based on the variables $\bar{S}_f(x)$ and $\bar{S}_m(y)$. In fact, substituting (5.20) into (5.19) and (5.21), we have

$$\begin{aligned} \bar{S}_f(x) &= (1 - \gamma) e^{-\lambda x} \pi_f(x) \\ &\quad - \int_0^{\omega} \int_0^{x \wedge y} e^{-\lambda\tau} \pi_c(x, y; \tau) \mathcal{M}(x - \tau, y - \tau; \bar{S}_f, \bar{S}_m) d\tau dy, \end{aligned} \tag{5.22}$$

$$\begin{aligned} \bar{S}_m(y) &= \gamma e^{-\lambda y} \pi_m(y) \\ &\quad - \int_0^{\omega} \int_0^{x \wedge y} e^{-\lambda\tau} \pi_c(x, y; \tau) \mathcal{M}(x - \tau, y - \tau; \bar{S}_f, \bar{S}_m) d\tau dx, \end{aligned} \tag{5.23}$$

$$\int_0^{\omega} \int_0^{\omega} \beta(x, y) \int_0^{x \wedge y} e^{-\lambda\tau} \pi_c(x, y; \tau) \mathcal{M}(x - \tau, y - \tau; \bar{S}_f, \bar{S}_m) d\tau dx dy = 1, \tag{5.24}$$

which—after some exchanges in the order of integration—may be written in the equivalent form

$$\begin{aligned} \bar{S}_f(x) &= (1 - \gamma) e^{-\lambda x} \pi_f(x) \\ &\quad - \int_0^{x} e^{-\lambda\tau} \int_{\tau}^{\omega} \pi_c(x, y; \tau) \mathcal{M}(x - \tau, y - \tau; \bar{S}_f, \bar{S}_m) dy d\tau, \end{aligned} \tag{5.25}$$

$$\begin{aligned} \bar{S}_m(y) &= \gamma e^{-\lambda y} \pi_m(y) \\ &\quad - \int_0^{y} e^{-\lambda\tau} \int_{\tau}^{\omega} \pi_c(x, y; \tau) \mathcal{M}(x - \tau, y - \tau; \bar{S}_f, \bar{S}_m) dx d\tau, \end{aligned} \tag{5.26}$$

$$\int_0^{\omega} e^{-\lambda\tau} \int_{\tau}^{\omega} \int_{\tau}^{\omega} \beta(x, y) \pi_c(x, y; \tau) \mathcal{M}(x - \tau, y - \tau; \bar{S}_f, \bar{S}_m) dx dy d\tau = 1. \tag{5.27}$$

With this formulation of the original problem (5.1)–(5.4) in just the variables $\bar{S}_f(x)$, $\bar{S}_m(y)$, and λ we are ready to discuss the existence of steady profiles—that is, of persistent solutions to the Fredrickson–Hoppensteadt model.

5.3 Existence of stationary profiles

The proof of existence of solutions to problem (5.25)–(5.27) is somewhat technical and is based on the classical Schauder fixed-point theorem. In addition to the assumption that the marriage function satisfies property (P3), we assume that (H1), (H2), and (H5) are also satisfied, together with the following additional hypothesis.

(H11) The female and male survival probabilities satisfy

$$\left|\frac{d}{dx}\pi_f(x)\right| = \mu_f(x)\pi_f(x) \le \pi^+, \qquad \left|\frac{d}{dy}\pi_m(y)\right| = \mu_m(y)\pi_m(y) \le \pi^+,$$

where π^+ is an appropriate positive constant.

Let us consider the variables $(u(x), v(y), \lambda)$ (where u and v are positive, bounded, continuous functions on the interval $[0, \omega)$ and λ is a real number) to construct a mapping that transforms such a triple $(u(x), v(y), \lambda)$ into a new triple $(\bar{u}(x), \bar{v}(y), \bar{\lambda})$ defined through

$$\bar{u}(x) = (1-\gamma)e^{-\lambda x}\pi_f(x) - \int_0^x K(x, \tau; u, v, \lambda)\bar{u}(\tau)d\tau, \tag{5.28}$$

$$\bar{v}(y) = \gamma e^{-\lambda y}\pi_m(y) - \int_0^y H(y, \tau; u, v, \lambda)\bar{v}(\tau)d\tau, \tag{5.29}$$

$$\bar{\lambda} = \left[\lambda - 1 + \int_0^\omega e^{-\lambda\tau}\int_\tau^\omega\int_\tau^\omega \beta(x, y)\pi_c(x, y; \tau)\mathcal{M}(x-\tau, y-\tau; u, v)dxdyd\tau\right]_{-\Lambda}^{\Lambda}. \tag{5.30}$$

The kernels K and H are, respectively, defined as

$$K(x, \tau; u, v, \lambda) = \frac{e^{-\lambda(x-\tau)}}{u(\tau)}\int_0^{\omega-(x-\tau)} \pi_c(x, y+x-\tau; x-\tau)\mathcal{M}(\tau, y; u, v)dy,$$

$$H(y, \tau; u, v, \lambda) = \frac{e^{-\lambda(y-\tau)}}{v(\tau)}\int_0^{\omega-(y-\tau)} \pi_c(x+y-\tau, y; y-\tau)\mathcal{M}(x, \tau; u, v)dx$$

for $0 \le \tau \le x < \omega$ and $0 \le \tau \le y < \omega$, respectively. Moreover, in (5.30) we have used the notation

$$[r]_{-\Lambda}^{\Lambda} = \begin{cases} -\Lambda & \text{if } r \le -\Lambda, \\ r & \text{if } -\Lambda \le r \le \Lambda, \\ \Lambda & \text{if } r \ge \Lambda, \end{cases}$$

where $\Lambda > 0$ is a fixed, arbitrarily chosen constant.

Note that, for a fixed triple $(u(x), v(y), \lambda)$, $\bar{u}(x)$, and $\bar{v}(y)$ are given as the solutions of the two uncoupled linear Volterra integral equations (5.28) and (5.29). Actually, for these equations we have the following result.

Theorem 5.1. *Let $\lambda \in \mathbb{R}$ and let $u(x)$, $v(y)$ be continuous functions on $[0, \omega]$ and positive on $[0, \omega)$. Then, (5.28) and (5.29) have unique solutions $\bar{u}(x)$ and $\bar{v}(y)$, continuous on $[0, \omega]$ and satisfying*

$$(1-\gamma)e^{-(\lambda+1)x}\pi_f(x) \le \bar{u}(x) \le (1-\gamma)e^{-\lambda x}\pi_f(x), \tag{5.31}$$

$$\gamma e^{-(\lambda+1)y}\pi_m(y) \le \bar{v}(y) \le \gamma e^{-\lambda y}\pi_m(y). \tag{5.32}$$

Proof. We focus only on (5.28), as (5.29) can be treated in exactly the same way.

Let us first set

$$z(x) = \frac{\bar{u}(x)e^{\lambda x}}{(1-\gamma)\pi_f(x)} \tag{5.33}$$

in (5.28) so that it simplifies into

$$z(x) = 1 - \int_0^x K_0(x, \tau; u, v)z(\tau)d\tau, \tag{5.34}$$

with

$$\begin{aligned} K_0(x, \tau; u, v) &= \frac{e^{\lambda(x-\tau)}\pi_f(\tau)}{\pi_f(x)} K(x, \tau; u, v, \lambda) \\ &= \frac{1}{u(\tau)} \int_0^{\omega-(x-\tau)} \pi_{m,\delta}(x, y + x - \tau; x - \tau)\mathcal{M}(\tau, y; u, v)dy, \end{aligned}$$

where we have used

$$\pi_{m,\delta}(x, y; z) = \frac{\pi_c(x, y; z)}{\pi_f(x; z)} = e^{-\int_0^z \mu_m(y-s)ds} e^{-\int_0^z \delta(x-s, y-s)ds}.$$

Now we note that, because of this definition and property (P4) of the marriage function, we have

$$0 \le K_0(x, \tau; u, v) \le \frac{1}{u(\tau)} \int_0^{\omega} \mathcal{M}(\tau, y; u, v)dy \le 1. \tag{5.35}$$

Moreover, we can see that K_0 is continuous in the set $0 \le \tau \le x < \omega$, so that (5.34) has a unique solution $z(x)$ that is continuous and bounded on $[0, \omega)$. Of course this implies that (see (5.33))

$$\bar{u}(x) = (1-\gamma)e^{-\lambda x}\pi_f(x)z(x)$$

is the unique solution to (5.28), and it can be continuously defined on $[0, \omega]$ (in particular $\bar{u}(\omega) = 0$).

To obtain a sharp estimate for $z(x)$ we need to observe that $K_0(x, \tau; u, v)$ is differentiable with respect to x in the region $0 \le \tau \le x < \omega$. In fact we have

$$\begin{aligned} &\frac{\partial}{\partial x} K_0(x, \tau; u, v) \\ &= -\frac{1}{u(\tau)} \int_0^{\omega-(x-\tau)} \mu_m(y + x - \tau)\pi_{m,\delta}(x, y + x - \tau; x - \tau)\mathcal{M}(\tau, y; u, v)dy \\ &\quad - \frac{1}{u(\tau)} \int_0^{\omega-(x-\tau)} \delta(x, y + x - \tau)\pi_{m,\delta}(x, y + x - \tau; x - \tau)\mathcal{M}(\tau, y; u, v)dy, \end{aligned} \tag{5.36}$$

where we have used $\pi_{m,\delta}(x, \omega; z) = 0$. We see that, thanks to assumption (H11) and to property (P4), the function $\frac{\partial}{\partial x} K_0(x, \tau; u, v)$ is nonpositive and bounded. Thus, from (5.34), we have that $z(x)$ is differentiable in $[0, \omega)$ and

$$z'(x) = -K_0(x, x; u, v)z(x) - \int_0^x \frac{\partial}{\partial x} K_0(x, \tau; u, v)z(\tau)d\tau, \tag{5.37}$$

so that

$$z(x) = e^{-\int_0^x K_0(\eta,\eta;u,v)d\eta}$$
$$- \int_0^x \int_\tau^x e^{-\int_s^x K_0(\eta,\eta;u,v)d\eta} \frac{\partial}{\partial s} K_0(s,\tau;u,v)ds\, z(\tau)d\tau. \tag{5.38}$$

Hence, we find that $z(x) \geq 0$. Furthermore, by (5.34) we have $z(x) \leq 1$ and, using (5.35) in (5.38), we also see that $z(x) \geq e^{-x}$. Of course these two inequalities for $z(x)$ prove (5.30). □

The previous theorem provides the basic step for the definition of our mapping, through (5.28)–(5.30), in the space $C([0,\omega]) \times C([0,\omega]) \times \mathbb{R}$. In particular, the estimates (5.31)–(5.32) imply that this mapping leaves the following closed convex set unchanged:

$$\begin{cases} (1-\gamma)e^{-(\Lambda+1)x}\pi_f(x) \leq u(x) \leq (1-\gamma)e^{\Lambda x}\pi_f(x), \\ \gamma e^{-(\Lambda+1)y}\pi_m(y) \leq v(y) \leq \gamma e^{\Lambda y}\pi_m(y), \\ -\Lambda \leq \lambda \leq \Lambda, \end{cases} \tag{5.39}$$

so that we may restrict the search for fixed points to this set. The following theorem states that the conditions for the existence of such a fixed point are satisfied.

Theorem 5.2. *The mapping* $(u,v,\lambda) \to (\bar{u},\bar{v},\bar{\lambda})$*, defined through* (5.28)–(5.30) *on the closed convex set* (5.39) *of the space* $C([0,\omega]) \times C([0,\omega]) \times \mathbb{R}$*, is continuous and relatively compact.*

Proof. We first prove continuity. Let (u_n, v_n, λ_n) be any sequence belonging to the set defined in (5.39) such that

$$(u_n, v_n, \lambda_n) \to (u,v,\lambda) \quad \text{in} \quad C([0,\omega]) \times C([0,\omega]) \times \mathbb{R}. \tag{5.40}$$

By property (P4) and assumption (H5) we have

$$|K(x,\tau;u_n,v_n,\lambda_n) - K(x,\tau;u,v,\lambda)|$$
$$\leq \frac{Q}{\pi_f(\tau)} \left(|u_n(\tau) - u(\tau)| + \|u_n - u\| + \|v_n - v\| + |\lambda_n - \lambda|\right),$$

where we have used the notation $\|f\| = \sup_{x\in[0,\omega]} |f(x)|$ and Q is a constant independent of n. Thus, the solutions of (5.28) satisfy

$$|\bar{u}_n(x) - \bar{u}(x)| \leq Q\left(\int_0^x |\bar{u}_n(\tau) - \bar{u}(\tau)|\,d\tau + \|u_n - u\| + \|v_n - v\| + |\lambda_n - \lambda|\right),$$

where Q is again a constant independent of n. Then, by Gronwall's lemma, we have

$$|\bar{u}_n(x) - \bar{u}(x)| \leq Q\left(\|u_n - u\| + \|v_n - v\| + |\lambda_n - \lambda|\right) \to 0$$

as $n \to \infty$. Finally, we can show in a similar way that

$$\bar{v}_n \to \bar{v} \quad \text{in} \quad C([0,\omega]), \quad \bar{\lambda}_n \to \bar{\lambda} \quad \text{in} \quad \mathbb{R},$$

proving that the mapping is continuous.

To prove relative compactness we note that if (u_n, v_n, λ_n) is a sequence satisfying (5.39), then

$$\bar{u}_n(x) = (1-\gamma)e^{-\lambda_n x}\pi_f(x)z_n(x),$$

where $z_n(x)$ satisfies

$$z_n(x) = 1 - \int_0^x K_0(x,\tau; u_n, v_n)z_n(\tau)d\tau.$$

Now, from the proof of Theorem 5.1 we know that z_n is bounded uniformly in n and differentiable on $[0,\omega)$, with

$$z_n'(x) = -K_0(x,x;u_n,v_n)z_n(x) - \int_0^x \frac{\partial}{\partial x}K_0(x,\tau;u_n,v_n)z_n(\tau)d\tau.$$

Since (see (5.36))

$$\left|\frac{\partial}{\partial x}K_0(x,\tau;u_n,v_n)\right| \le Q,$$

where the constant Q does not depend on n, we have that $\bar{u}_n(x)$ is differentiable on $[0,\omega)$ and the sequence of the derivatives is bounded uniformly in n. The same holds for the sequence $\bar{v}_n(y)$, so that the mapping is relatively compact. □

On the basis of the results in Theorem 5.1, we obtain existence of a fixed point for the mapping defined by (5.28)–(5.30). We call (u^*, v^*, λ^*) such a fixed point that is a candidate for being a solution to the original problem (5.25)–(5.27). We have to make sure that λ^* lies in the interior of the interval $[-\Lambda, \Lambda]$. For this purpose it is enough to take Λ sufficiently large. In fact if we set

$$\mathcal{R}_+(\lambda) = \sup_{(u,v)\in\mathcal{S}_+} \int_0^\omega e^{-\lambda\tau}\int_\tau^\omega\int_\tau^\omega \beta(x,y)\pi_c(x,y;\tau)\mathcal{M}(x-\tau,y-\tau;u,v)dxdyd\tau,$$

$$\mathcal{R}_-(\lambda) = \inf_{(u,v)\in\mathcal{S}_-} \int_0^\omega e^{-\lambda\tau}\int_\tau^\omega\int_\tau^\omega \beta(x,y)\pi_c(x,y;\tau)\mathcal{M}(x-\tau,y-\tau;u,v)dxdyd\tau,$$

$$m_-(x,y) = \inf_{(u,v)\in\mathcal{S}_-} \mathcal{M}(x,y;u,v),$$

where

$$\mathcal{S}_+ = \left\{(u,v)|\ u(x) \le (1-\gamma)\pi_f(x)\ ;\ v(y) \le \gamma\pi_m(y)\right\},$$

$$\mathcal{S}_- = \left\{(u,v)|\ u(x) \ge (1-\gamma)e^{-x}\pi_f(x)\ ;\ v(y) \ge \gamma e^{-y}\pi_m(y)\right\},$$

we have the following result.

Lemma 5.3.

$$\lim_{\lambda\to+\infty} \mathcal{R}_+(\lambda) = 0. \tag{5.41}$$

If also

$$\int_0^\omega \int_0^\omega \beta(x, y)m_-(x, y)dxdy > 0, \tag{5.42}$$

then

$$\lim_{\lambda\to-\infty} \mathcal{R}_-(\lambda) = +\infty. \tag{5.43}$$

Proof. The proof of (5.41) is straightforward, since, for $(u, v) \in \mathcal{S}_+$, we have (using (P4))

$$\int_0^\omega e^{-\lambda\tau} \int_\tau^\omega \int_\tau^\omega \beta(x, y)\pi_c(x, y; \tau)\mathcal{M}(x-\tau, y-\tau; u, v)dxdyd\tau \\ \leq \bar{\beta}(1-\gamma)\omega \int_0^\omega e^{-\lambda\tau} d\tau.$$

To prove (5.43) we note that, if (5.42) is satisfied, then there exists $T > 0$ small enough that, for $\tau \in [0, T]$, $(u, v) \in \mathcal{S}_-$, and m sufficiently small, we have

$$\int_\tau^\omega \int_\tau^\omega \beta(x, y)\pi_c(x, y; \tau)\mathcal{M}(x-\tau, y-\tau; u, v)dxdy \\ \geq \int_\tau^\omega \int_\tau^\omega \beta(x, y)\pi_c(x, y; \tau)m_-(x-\tau, y-\tau)dxdy \geq m > 0.$$

Then,

$$\int_0^\omega e^{-\lambda\tau} \int_\tau^\omega \int_\tau^\omega \beta(x, y)\pi_c(x, y; \tau)\mathcal{M}(x-\tau, y-\tau; u, v)dxdyd\tau \geq m \int_0^T e^{-\lambda\tau} d\tau$$

and (5.42) holds. □

Thus we finally arrive at the following result.

Theorem 5.4. *Let the assumptions* (H1), (H2), (H5), (H11) *hold and suppose that the marriage function satisfies* (P3) *and condition* (5.42). *Then, problem* (5.25)–(5.27) *has at least one solution* (u^*, v^*, λ^*). *Moreover, if*

$$\mathcal{R}_+(0) < 1, \tag{5.44}$$

then $\lambda^* < 0$. *If instead*

$$\mathcal{R}_-(0) > 1, \tag{5.45}$$

then $\lambda^* > 0$.

Proof. Consider the mapping defined through (5.28)–(5.30), where Λ is sufficiently large that

$$\mathcal{R}_+(\lambda) < 1 \quad \text{for} \quad \lambda \geq \Lambda,$$

$$\mathcal{R}_-(\lambda) > 1 \quad \text{for} \quad \lambda \leq -\Lambda.$$

Let (u^*, v^*, λ^*) be a fixed point whose existence is proved in Theorem 5.2. We must show that $\lambda^* \in (-\Lambda, \Lambda)$. We first prove that $\lambda^* \neq \Lambda$. In fact if $\lambda^* = \Lambda$, we have (see (5.39))

$$u^*(x) \leq (1-\gamma)e^{-\Lambda x}\pi_f(x) \leq (1-\gamma)\pi_f(x),$$

$$v^*(y) \leq \gamma e^{-\Lambda y}\pi_m(y) \leq \gamma\pi_m(y),$$

so that $(u^*, v^*) \in \mathcal{S}_+$. Then,

$$\int_0^\omega e^{-\Lambda\tau}\int_\tau^\omega\int_\tau^\omega \beta(x,y)\pi_c(x,y;\tau)\mathcal{M}(x-\tau, y-\tau; u^*, v^*)dxdyd\tau \leq \mathcal{R}_+(\Lambda) < 1,$$

so that

$$\Lambda = \left[\Lambda - 1 + \int_0^\omega e^{-\Lambda\tau}\int_\tau^\omega\int_\tau^\omega \beta(x,y)\pi_c(x,y;\tau)\mathcal{M}(x-\tau, y-\tau; u^*, v^*)dxdyd\tau\right]_{-\Lambda}^{\Lambda} < \Lambda,$$

a contradiction. In a similar way we can prove that $\lambda^* \neq -\Lambda$. Finally, we note that, if $\lambda^* > 0$, then

$$1 = \int_0^\omega e^{-\lambda^*\tau}\int_\tau^\omega\int_\tau^\omega \beta(x,y)\pi_c(x,y;\tau)\mathcal{M}(x-\tau, y-\tau; u^*, v^*)dxdyd\tau < \mathcal{R}_+(0),$$

a contradiction if (5.44) is satisfied. In a similar way we can prove that if (5.45) holds, then it is impossible that $\lambda^* < 0$. □

The previous theorem is the main result we can obtain under the general assumptions we made on the marriage function. We note that we need condition (5.42) in order to deal with possibly negative λ's. This additional condition is natural, however, since it states that the population has a nontrivial birth rate even at the "minimal" marriage rate. Thus, trivial situations such as that in which the marriage function produces only couples that are not fertile are avoided. We shall say that if (5.42) is satisfied, then the population is *reproductive*. If we consider the marriage function given by the harmonic mean (E3), for example, we have

$$m_-(x,y) = 2\rho(x,y)\frac{\gamma(1-\gamma)e^{-x}\pi_f(x)e^{-y}\pi_m(y)}{(1-\gamma)e^{-x}\pi_f(x) + \gamma e^{-y}\pi_m(y)},$$

and condition (5.42) is satisfied if the product $\beta(x,y)\rho(x,y)$ does not vanish identically.

5.4 Stability of stationary profiles

The main demographic implication of the one-sex stable population theory is the result describing the long-term behavior of solutions. Namely, though the population as a whole undergoes exponential growth (or decay), its age profile attains a uniquely defined profile. For the two-sex model we found that multiple stationary profiles are possible, so that we are led to analyze the stability of each of these profiles with respect to local perturbations. We perform this analysis via the standard technique of linearization applied to system (5.1)–(5.4). For this purpose we keep all the assumptions (H1), (H2), (H5), (H11), and (P3) holding in section 5.3 and, moreover, we consider the following one connected with assumption (H5).

(H12) The basic function $\mathcal{F}[x, y, f, m]$ defining the marriage function of the first kind in assumption (H5) is differentiable with respect to the variables f, m at any point different from $f = m = 0$.

Then, we consider a stationary profile $(\mathcal{U}_f^*(x), \mathcal{U}_m^*(y), \mathcal{C}^*(x, y), \lambda^*)$, solution of (5.8)–(5.11), together with the consequent stationary distribution for the singles:

$$\mathcal{S}_f^*(x) = \mathcal{U}_f^*(x) - \int_0^\omega \mathcal{C}^*(x, y)dy, \quad \mathcal{S}_m^*(y) = \mathcal{U}_m^*(y) - \int_0^\omega \mathcal{C}^*(x, y)dx.$$

Next, we take the small deviations

$$\begin{aligned}
&\delta_f(x, t) = \mathcal{U}_f(x, t) - \mathcal{U}_f^*(x),\\
&\delta_m(y, t) = \mathcal{U}_m(y, t) - \mathcal{U}_m^*(y),\\
&\delta_c(x, y, t) = \mathcal{C}(x, y, t) - \mathcal{C}^*(x, y),\\
&\delta_\lambda(t) = \lambda(t) - \lambda^*,\\
&\delta_f^s(x, t) = \mathcal{S}_f(x, t) - \mathcal{S}_f^*(x) = \delta_f(x, t) - \int_0^\omega \delta_c(x, y, t)dy,\\
&\delta_m^s(y, t) = \mathcal{S}_m(y, t) - \mathcal{S}_m^*(y) = \delta_m(y, t) - \int_0^\omega \delta_c(x, y, t)dx.
\end{aligned}$$

These deviations approximately satisfy the following linearized version of (5.1)–(5.4):

$$\begin{cases}
\dfrac{\partial \delta_f}{\partial t} + \dfrac{\partial \delta_f}{\partial x} = -\lambda^* \delta_f(x, t) - \mu_f(x)\delta_f(x, t) + \delta_\lambda(t)\mathcal{U}_f^*(x),\\
\delta_f(0, t) = \displaystyle\int_0^\omega \int_0^\omega \beta_f(x, y)\delta_c(x, y, t)dxdy,
\end{cases} \tag{5.46}$$

$$\begin{cases}
\dfrac{\partial \delta_m}{\partial t} + \dfrac{\partial \delta_m}{\partial y} = -\lambda^* \delta_m(y, t) - \mu_m(y)\delta_m(y, t) + \delta_\lambda(t)\mathcal{U}_m^*(y),\\
\delta_m(0, t) = \displaystyle\int_0^\omega \int_0^\omega \beta_m(x, y)\delta_c(x, y, t)dxdy,
\end{cases} \tag{5.47}$$

$$\begin{cases}
\dfrac{\partial \delta_c}{\partial t} + \dfrac{\partial \delta_c}{\partial x} + \dfrac{\partial \delta_c}{\partial y} = -\lambda^* \delta_c(x, y, t) - \sigma(x, y)\delta_c(x, y, t) + \delta_\lambda(t)\mathcal{C}^*(x, y)\\
+\dfrac{\partial \mathcal{F}}{\partial f}\left[x, y, \mathcal{S}_f^*(x), \mathcal{S}_m^*(y)\right]\delta_f^s(x, t) + \dfrac{\partial \mathcal{F}}{\partial m}\left[x, y, \mathcal{S}_f^*(x), \mathcal{S}_m^*(y)\right]\delta_m^s(y, t),\\
\delta_c(x, 0, t) = \delta_c(0, y, t) = 0,
\end{cases} \tag{5.48}$$

with the constraint

$$\int_0^\omega \delta_f(x, t)dx + \int_0^\omega \delta_m(y, t)dy = 0. \tag{5.49}$$

Now, the procedure to determine stability of the stationary states consists of finding exponential solutions of (5.46)–(5.49) in the form

$$\begin{aligned}
\delta_f(x,t) &= e^{kt}\Delta_f(x),\\
\delta_m(y,t) &= e^{kt}\Delta_m(y),\\
\delta_c(x,y,t) &= e^{kt}\Delta_c(x,y),\\
\delta_\lambda(t) &= e^{kt}\Delta_\lambda,
\end{aligned}$$

that is, to solve the eigenvalue problem

$$\begin{cases}
k\Delta_f + \dfrac{\partial \Delta_f}{\partial x} = -\lambda^*\Delta_f(x) - \mu_f(x)\Delta_f(x) + \Delta_\lambda \mathcal{U}_f^*(x),\\[2ex]
\Delta_f(0) = \displaystyle\int_0^\omega\int_0^\omega \beta_f(x,y)\Delta_c(x,y)dxdy,
\end{cases} \tag{5.50}$$

$$\begin{cases}
k\Delta_m + \dfrac{\partial \Delta_m}{\partial y} = -\lambda^*\Delta_m(y) - \mu_m(y)\Delta_m(y) + \Delta_\lambda \mathcal{U}_m^*(y),\\[2ex]
\Delta_m(0) = \displaystyle\int_0^\omega\int_0^\omega \beta_m(x,y)\Delta_c(x,y)dxdy,
\end{cases} \tag{5.51}$$

$$\begin{cases}
k\Delta_c + \dfrac{\partial \Delta_c}{\partial x} + \dfrac{\partial \Delta_c}{\partial y} = -\lambda^*\Delta_c(x,y) - \sigma(x,y)\Delta_c(x,y) + \Delta_\lambda \mathcal{C}^*(x,y)\\[2ex]
\quad + \dfrac{\partial \mathcal{F}}{\partial f}\left[x,y,\mathcal{S}_f^*(x),\mathcal{S}_m^*(y)\right]\Delta_f^s(x) + \dfrac{\partial \mathcal{F}}{\partial m}\left[x,y,\mathcal{S}_f^*(x),\mathcal{S}_m^*(y)\right]\Delta_m^s(y),\\[2ex]
\Delta_c(x,0) = \Delta_c(0,y) = 0,
\end{cases} \tag{5.52}$$

$$\int_0^\omega \Delta_f(x)dx + \int_0^\omega \Delta_m(y)dy = 0, \tag{5.53}$$

where

$$\Delta_f^s(x) = \Delta_f(x) - \int_0^\omega \Delta_c(x,y)dy, \quad \Delta_m^s(y) = \Delta_m(y) - \int_0^\omega \Delta_c(x,y)dx.$$

Once we find the eigenvalues then we apply the following criterion.

Theorem 5.5. *If all eigenvalues k of the eigenvalue problem* (5.50)–(5.53) *are such that*

$$\Re k < 0,$$

then the stationary profile is stable. If there is one eigenvalue $\bar{k}$ such that

$$\Re \bar{k} > 0,$$

then the steady state is unstable.

5.5 Some examples

In this section we consider some particular cases that allow some elaboration in order to state existence and uniqueness of a stationary profile. The previous theory is, in fact, too general to provide precise knowledge of the situation. However, we show some specific cases that can be treated completely.

The first example concerns the simplest form for the marriage function, that is, the *female dominance* case (E1):

$$\mathcal{M}(x, y, t; s_f, s_m) = \rho(x, y)s_f(x, t),$$

where

$$\int_0^\omega \rho(x, y)dy \leq 1.$$

With these choices (5.25) becomes

$$\bar{\mathcal{S}}_f(x) = (1-\gamma)e^{-\lambda x}\pi_f(x) - \int_0^x e^{-\lambda\tau}\frac{\pi_f(x)}{\pi_f(x-\tau)}K(x,\tau)\bar{\mathcal{S}}_f(x-\tau)d\tau,$$

where

$$K(x,\tau) = \int_\tau^\omega \pi_{m,\delta}(x, y; \tau)\rho(x-\tau, y)dy \leq 1.$$

Now, setting

$$u(x) = \frac{e^{\lambda x}\bar{\mathcal{S}}_f(x)}{\pi_f(x)},$$

we have the linear Volterra equation

$$u(x) = (1-\gamma) - \int_0^x K(x,\tau)u(x-\tau)d\tau,$$

which has a unique solution we call $\Sigma(x)$ and does not depend on λ. Note that $\Sigma(x)$ is nonnegative, as can be seen from the proof of Theorem 5.1. Thus, the density of single females has the form

$$\bar{\mathcal{S}}_f(x) = e^{-\lambda x}\pi_f(x)\Sigma(x),$$

and we only have to determine λ. We use this expression for $\bar{\mathcal{S}}_f(x)$ in (5.27) and obtain the following equation in λ:

$$\int_0^\omega\int_\tau^\omega\int_\tau^\omega e^{-\lambda x}\beta(x, y)\rho(x-\tau, y-\tau)\pi_f(x)\pi_{m,\delta}(x, y; \tau)\Sigma(x-\tau)dxdyd\tau = 1.$$

We see that, since the left-hand side is decreasing with λ (if we exclude the trivial case in which it is constantly zero) and has limits 0 and $+\infty$, respectively, as λ goes to $+\infty$ and $-\infty$, this equation has one and only one real solution λ^* that is positive provided the following threshold condition is satisfied:

$$\int_0^\omega\int_\tau^\omega\int_\tau^\omega \beta(x, y)\rho(x-\tau, y-\tau)\pi_f(x)\pi_{m,\delta}(x, y; \tau)\Sigma(x-\tau)dxdyd\tau > 1.$$

Thus, the steady profile for the single females is

$$\bar{\mathcal{S}}_f(x) = e^{-\lambda^* x}\pi_f(x)\Sigma(x), \tag{5.54}$$

and the corresponding profile $\bar{\mathcal{S}}_m(y)$ for the males can be obtained by substituting (5.54) into (5.26). In the same way we can get from (5.20) the profile for the couples. We finally note that once we have found λ^* we may go back to the original formulation of the problem (5.8)–(5.11) and find the following expressions for the profiles:

$$\mathcal{U}_f^*(x) = \frac{(1-\gamma)e^{-\lambda^* x}\pi_f(x)}{\int_0^\omega e^{-\lambda^* s}\left[(1-\gamma)\pi_f(s) + \gamma\pi_m(s)\right]ds},$$

$$\mathcal{U}_m^*(y) = \frac{\gamma e^{-\lambda^* y}\pi_m(y)}{\int_0^\omega e^{-\lambda^* s}\left[(1-\gamma)\pi_f(s) + \gamma\pi_m(s)\right]ds},$$

$$\mathcal{C}^*(x, y) = \frac{1}{1-\gamma}\,\mathcal{U}_f^*(x)\int_0^{x\wedge y} \pi_{m,\delta}(x, y; \tau)\rho(x-\tau, y-\tau)\Sigma(x-\tau)d\tau.$$

As a second example we consider the marriage function

$$\mathcal{M}(x, y, t; s_f, s_m) = \rho\frac{s_f(x, t)s_m(y, t)}{\int_0^\infty s_f(\tau)d\tau + \int_0^\infty s_m(\tau)d\tau} \tag{5.55}$$

together with the assumptions

$$\omega = \infty, \quad \mu_f(\tau) \equiv \mu_m(\tau) \equiv \mu, \quad \delta(x, y) \equiv 0, \quad \gamma = \frac{1}{2}, \quad \beta(x, y) \equiv \beta. \tag{5.56}$$

The marriage function (5.55) is of the second kind and the assumptions (5.56) help to simplify the problem by reducing it to an algebraic computation. Actually, in solving (5.25)–(5.27) we shall use the scalar variables

$$F = \int_0^\infty \bar{\mathcal{S}}_f(\tau)\,d\tau, \quad M = \int_0^\infty \bar{\mathcal{S}}_m(\tau)\,d\tau.$$

Then, integrating (5.25)–(5.26) and using (5.56), we get, for $\lambda > -\mu$,

$$F = \frac{1}{2(\lambda+\mu)} - \frac{\rho F M}{(\lambda+2\mu)(F+M)},$$

$$M = \frac{1}{2(\lambda+\mu)} - \frac{\rho F M}{(\lambda+2\mu)(F+M)},$$

so that we see $F = M$. We can explicitly compute their value,

$$F = \frac{\lambda+2\mu}{2(\lambda+\mu)(2\lambda+4\mu+\rho)} \tag{5.57}$$

and, putting $F = M$ in (5.25), we derive the equation

$$\bar{\mathcal{S}}_f(x) = \frac{1}{2}e^{-(\lambda+\mu)x} - \frac{\rho}{2}\int_0^x e^{-(\lambda+2\mu)(x-\tau)}\bar{\mathcal{S}}_f(\tau)d\tau.$$

This can be explicitly solved, yielding

$$\bar{S}_f(x) = \frac{\rho}{2(\rho+2\mu)} e^{-(\lambda+2\mu+\frac{\rho}{2})x} + \frac{\mu}{\rho+2\mu} e^{-(\lambda+\mu)x}, \tag{5.58}$$

and the same expression is obtained for $\bar{S}_m(y)$. To determine λ^* we use (5.57) and (5.58) in (5.27) to get

$$(\lambda+\mu)(2\lambda+4\mu+\rho) = \frac{\rho\beta}{4},$$

which has one and only one solution $\lambda^* > -\mu$. This solution is positive if the following condition is satisfied:

$$\rho\beta - 4\mu(4\mu+\rho) > 0.$$

5.6 A numerical example of a persistent two-sex population

In trying to understand the structure of two-sex populations at steady state we were able to derive explicit formulas for the age densities of females and males in terms of the (unknown) Malthusian rate λ and the probabilities of survival π_f and π_m. These steady states are given by

$$\begin{cases} u_f(x) = (1-\gamma)\, e^{-\lambda x}\, \pi_f(x), \\ \\ u_m(y) = \gamma\, e^{-\lambda y}\, \pi_m(y), \end{cases} \tag{5.59}$$

where γ is the proportion of boys among all newborns. However, no explicit form for the density of couples can be found in general, but only in extremely simplified situations (see section 5.5).

In this section we shall present the results of a numerical simulation performed under the assumption that, after a long time, the population densities will be represented by their steady states multiplied by an exponential factor in time:

$$\begin{cases} u_f(x,t) = (1-\gamma)\, e^{\lambda t}\, e^{-\lambda x}\, \pi_f(x), \\ \\ u_m(y,t) = \gamma\, e^{\lambda t}\, e^{-\lambda y}\, \pi_m(y), \\ \\ c(x,y,t) = e^{\lambda t}\, \bar{c}(x,y). \end{cases} \tag{5.60}$$

We remark that both λ and $\bar{c}$ are unknown and must be determined from the simulation. The total population is given by summing the female and male densities over all ages and adding the results:

$$P(t) = e^{\lambda t}\left[(1-\gamma)\int_0^\omega e^{-\lambda x}\,\pi_f(x)\,dx + \gamma\int_0^\omega e^{-\lambda y}\,\pi_m(y)\,dy\right]. \tag{5.61}$$

The simulator computes approximations X_i^n of $u_f(x_i, t^n)$ and Y_j^n of $u_m(y_j, t^n)$, from which an approximation $\tilde{P}^n$ of $P^n = P(t^n)$ is obtained by quadrature using the composite Simpson's rule. Then, the simulator evaluates the approximate rate of increase of the total population from each time to the next from the relation $\tilde{P}^n/\tilde{P}^{n-1} = e^{\lambda^n(t^n - t^{n-1})}$:

$$\lambda^n = \frac{\ln(\tilde{P}^n/\tilde{P}^{n-1})}{h}. \tag{5.62}$$

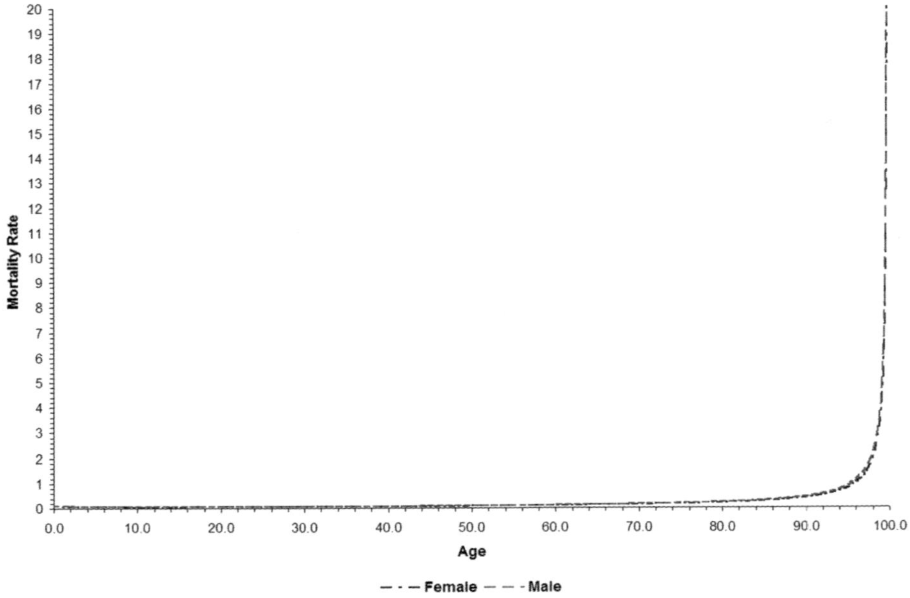

Figure 5.1. *Mortality rates.*

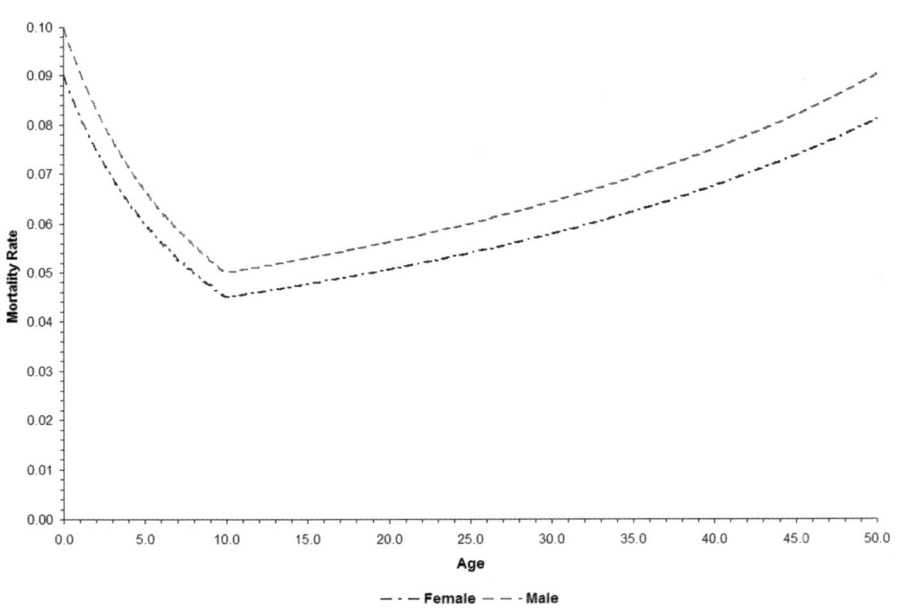

Figure 5.2. *Mortality rates to 50 years of age.*

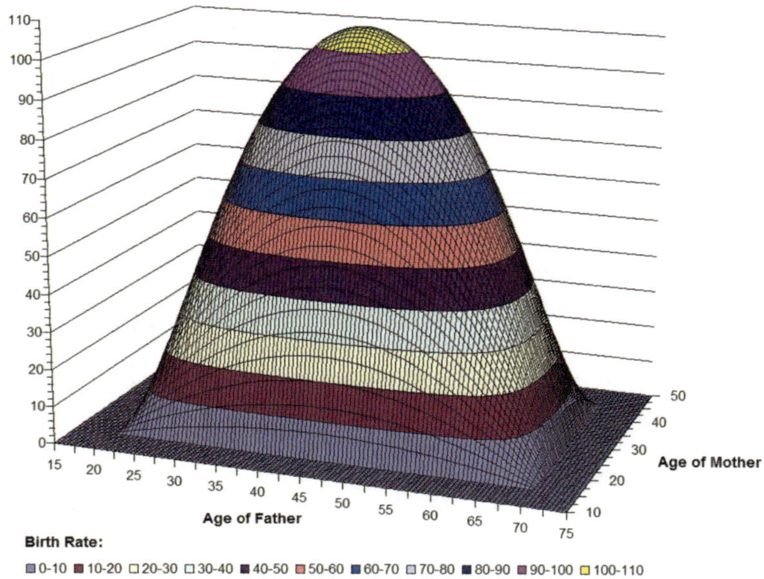

Figure 5.3. *Birth rate.*

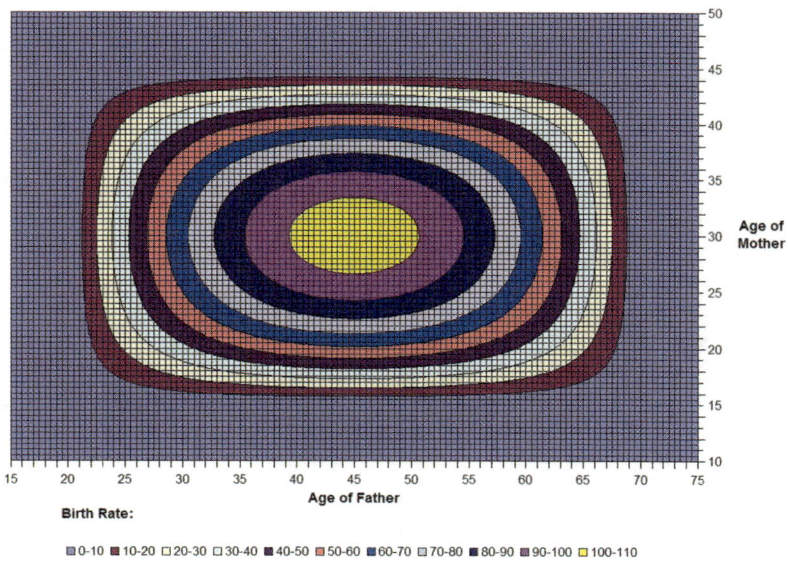

Figure 5.4. *Contour plot of birth rate.*

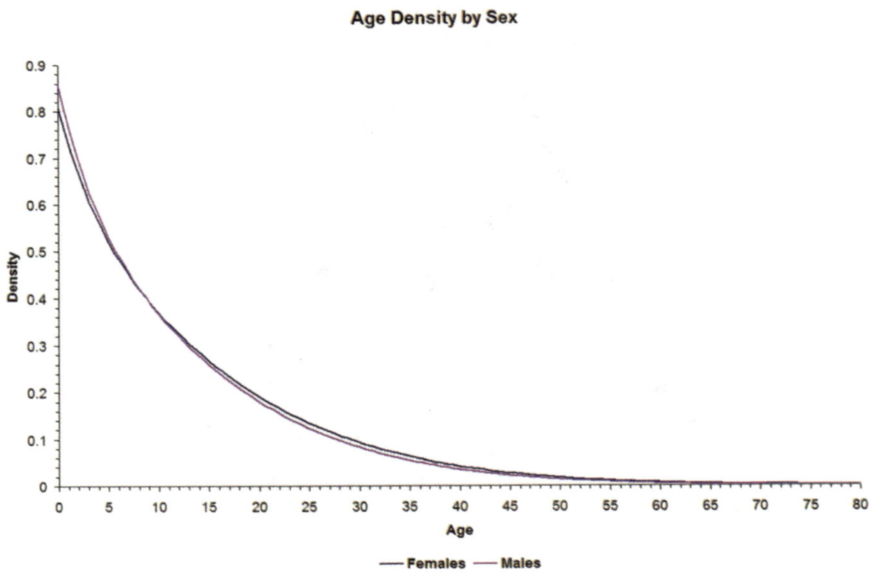

Figure 5.5. *Densities of females and males.*

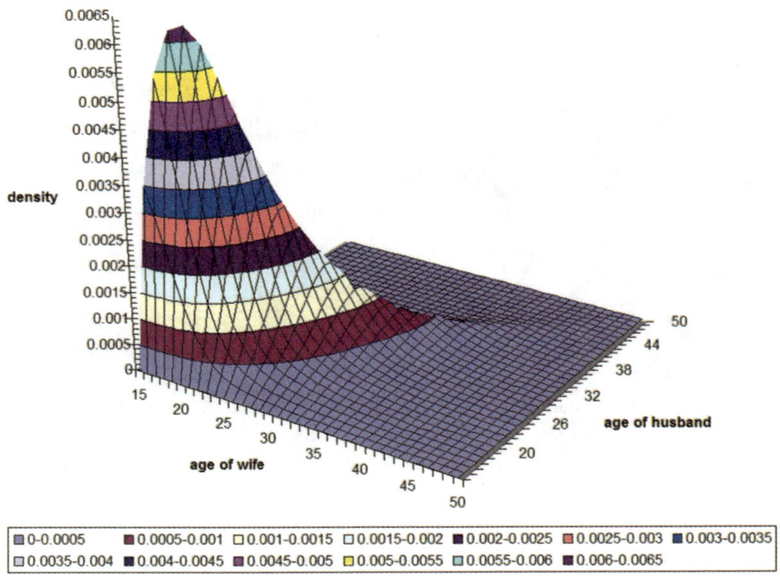

Figure 5.6. *Density of couples.*

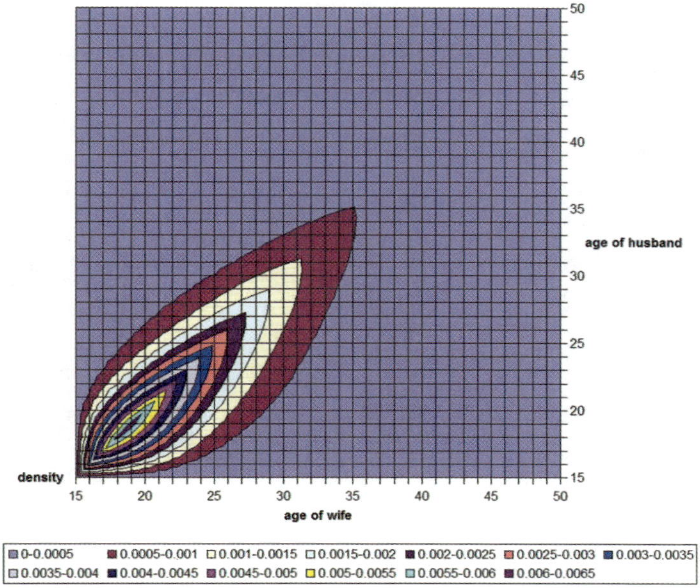

Figure 5.7. *Contour plot of density of couples.*

When successive values of λ^n are unchanged for many time-steps, it is assumed that this value is the Malthusian rate λ and the resulting distribution of couples is stored in a file. Finally, the steady-state distribution of couples is obtained from the latter through multiplication by $e^{-\lambda t^n}$, and distributions for females and males are obtained directly from (5.60) using the value of λ just found.

We initialize the population with $\lambda = 0.02$, the female and male age densities given by (5.59), and the distribution of couples—arbitrarily—by

$$Z_{ij}^0 = \frac{(1-\gamma)\,\pi_f(x_i)\,\gamma\,\pi_m(y_j)}{F_0 M_0},$$

where F_0 and M_0 are, respectively, the total initial numbers of females and males.

We show in Figures 5.1 and 5.2 the graph of the mortality rates we used.

Also, in Figures 5.3 and 5.4, we show the graph of the birth rate we used, respectively, as a three-dimensional surface and a contour plot.

For the dissolution rate of couples we used $\sigma(x, y) = \mu_f(x) + \mu_m(y) + 0.01$, and for the marriage function the harmonic mean (E3), $\mathcal{M}(x, y; s_f, s_m) = 2\rho(x, y)\,\frac{s_f\, s_m}{s_f + s_m}$, with

$$\rho(x, y) = \begin{cases} 0.0005 & \text{if } x > 15 \text{ and } y > 15, \\ 0 & \text{otherwise.} \end{cases}$$

The Malthusian rate we obtain is $\lambda \approx 0.017621$, and the resulting densities of females, males, and couples are presented in Figures 5.5, 5.6, and 5.7.

Appendix

The Main Algorithm

In this appendix we provide a FORTRAN code for the algorithm based on the numerical method described and analyzed in the previous chapters.

Several files with parameter input data are needed to run a simulation, and we shall describe them in detail. First, a file containing thirty-five lines is needed: the first nine lines being numbers and the remaining being twenty-six filenames. Its first three lines contain minimum values for the separation rates, death rates, and other rates—respectively, 0.001, 0.0001, and 0.0001. The next two lines contain positive integers that indicate, respectively, the number of times during a simulation that output should be written to files and the total number of time-steps the simulation should run. The next line contains the step size, and the following two lines the year for the first set of vital statistics data and the year for the second set of vital statistics data. The ninth line has a -1.0 if the linear-in-time data is interpolated from an earlier time, a 1.0 if it is extrapolated from a later time, and a 0.0 if the rates are constant in time. The final twenty-six lines contain the names of the files in which the vital and demographic rates for the simulation are stored.

Lines 10–15 contain the names of the data files in which data for distributions of males, females, and couples are stored: lines 10 and 11 contain, respectively, the names of the files for the distribution of males for the first year for which data are stored and for the distribution of males for the second year. The next two lines contain the same kind of data for females, and the following two for couples. The next four lines (16–19) contain the names of the files with the numbers of marriage annulments by age of one spouse—the first two for males and the next two for females, first for the first year, then for the last year. The next eight lines (20–27) contain the names of the files with natality data: four pairs, first for the first year and then for the second year. The first two contain birth rates for all women by age group. The next two contain birth rates for unmarried women by age group. The next two files contain the numbers of newborn boys by age of mother and by age of father. The final two files contain the numbers of newborn girls by age of mother and by age of father. Next come four lines (28–31) with the names of the files containing mortality data: the first two contain mortality rates for males in the first year and in the last year; the next two lines name the files containing mortality rates for females in the first year and in the last year. Finally, lines 32–35 contain the names of the files where marriage data are stored: the first two contain marriage rates for males in the first year and in the last year; the next two contain marriage rates for females in the first year and in the last year. Here is an example of a file named "data" that contains the information just described.

```
.001
.0001
.0001
1
160
0.0625
1980
1990
-1.0
cenm80w5
cenm90
cenf80w5
cenf90
cenc80
cenc90
vsannm80
vsannm90
vsannf80
vsannf90
vsbrrc80
vsbrrc90
vsbrrs80
vsbrrs90
vsbirm80
vsbirm90
vsbirf80
vsbirf90
vsdthm80
vsdthm90
vsdthf80
vsdthf90
vsmarm80
vsmarm90
vsmarf80
vsmarf90
```

A.1 A FORTRAN algorithm

Here we include a FORTRAN algorithm based on the numerical method of the previous sections, which we use for projections of the population of the United States.

Algorithm A.1.

```
c
                    program agesex
c
c  this program sets up the functions for and calls the routine twosex
c
c------------------------------------------------------------------------
c
                    implicit real (a-h,o-z)
                    character*20 cenm(2), cenf(2), cenc(2),
                $               vsannm(2), vsannf(2),
                $               vsbirm(2), vsbirf(2), vsbrrc(2), vsbrrs(2),
                $               vsdthm(2), vsdthf(2),
                $               vsnmar(2),
                $               routin
                    integer time1, time2, tim
                    common/init/ time1, time2, extrap, maxdim, ierror
                    common/um0com/ num0, um0wid, um01(0:25), um02(0:25)
                    common/uf0com/ nuf0, uf0wid, uf01(0:25), uf02(0:25)
                    common/c0com/ nc0, c0wid, c01(0:25,0:25), c02(0:25,0:25)
                    common/anncom/ nann, annwid,
                $               annri(0:25,0:25), annrs(0:25,0:25)
                    common/bircom/ nbir, birwid
                    common/brcmcm/ brcmri(0:25,0:25), brcmrs(0:25,0:25)
                    common/brcfcm/ brcfri(0:25,0:25), brcfrs(0:25,0:25)
                    common/brsmcm/ brsmri(0:25), brsmrs(0:25)
                    common/brsfcm/ brsfri(0:25), brsfrs(0:25)
                    common/dthmcm/ ndthm, dwidm0, dwdsm1, dthwdm,
                $               dthrim(0:25), dthrsm(0:25)
                    common/dthfcm/ ndthf, dwidf0, dwdsf1, dthwdf,
                $               dthrif(0:25), dthrsf(0:25)
                    common/marcom/ nmar, marwid,
                $               nmari(0:25,0:25), nmars(0:25,0:25)
                    common/funcg/ gi(0:25,0:25), gs(0:25,0:25)
                    common/funch/ hi(0:25,0:25), hs(0:25,0:25)
                    common/temp1/ a1(0:25), a2(0:25)
                    common/temp2/ b1(0:25), b2(0:25)
                    common/mincom/ annmin, dthmin
                    common/rnderr/ zero
                    data  maxdim /25/
c
c  input initial information and data file names
c
                    ierror = 0
```

```
                 read(5,1020) annmin, dthmin, zero
                 read(5,1030) nprint, nstep, deltat
                 read(5,1000) time1, time2, extrap
                 read(5,1010) cenm(1), cenm(2), cenf(1), cenf(2),
     $             cenc(1), cenc(2),
     $             vsannm(1), vsannm(2), vsannf(1), vsannf(2),
     $             vsbrrc(1), vsbrrc(2), vsbrrs(1), vsbrrs(2),
     $             vsbirm(1), vsbirm(2), vsbirf(1), vsbirf(2),
     $             vsdthm(1), vsdthm(2), vsdthf(1), vsdthf(2),
     $             vsnmar(1), vsnmar(2)
 1000            format(i16,/,i16,/,g16.8)
 1010            format(a)
 1020            format(g16.8)
 1030            format(i16,/,i16,/,g16.8)
                 yrgap = time1 - time2
                 if((extrap * yrgap).lt.0.0) then
                     tim   = time2
                     time2 = time1
                     time1 = tim
                     i     = 2
                     j     = 1
                   else
                     i     = 1
                     j     = 2
                   endif
                 yrgap  = abs(yrgap)
                 if(yrgap.eq.0) then
                     extrap = 0.0
                   else
                     if(extrap.gt.0.0) extrap =  1.0 / yrgap
                     if(extrap.lt.0.0) extrap = -1.0 / yrgap
                     if(extrap.eq.0.0) then
                       j = 1
                       time2 = time1
                       endif
                   endif
c
c  read data files
c
                 open(unit=11,  file=cenm(i))
                 open(unit=12,  file=cenm(j))
                 call iu0 (11, 12, num0, um0wid, um01, um02)
                 if(ierror.ne.0) then
                   routin = 'iu0 males'
                   goto 9999
                   endif
```

```
      close(11)
      close(12)
c
      open(unit=21,  file=cenf(i))
      open(unit=22,  file=cenf(j))
      call iu0 (21, 22, nuf0, uf0wid, uf01, uf02)
      if(ierror.ne.0) then
        routin = 'iu0 females'
        goto 9999
        endif
      close(21)
      close(22)
c
      open(unit=31,  file=cenc(i))
      open(unit=32,  file=cenc(j))
      call ic0 (31, 32,  nc0,  c0wid,  c01,  c02)
      if(ierror.ne.0) then
        routin = 'ic0'
        goto 9999
        endif
      close(31)
      close(32)
c
      open(unit=41,  file=vsannm(i))
      open(unit=42,  file=vsannm(j))
      open(unit=43,  file=vsannf(i))
      open(unit=44,  file=vsannf(j))
      call iann(41, 42, 43, 44)
      if(ierror.ne.0) then
        routin = 'iann males'
        goto 9999
        endif
      close(41)
      close(42)
      close(43)
      close(44)
c
      open(unit=51,  file=vsbrrc(i))
      open(unit=52,  file=vsbrrc(j))
      open(unit=61,  file=vsbrrs(i))
      open(unit=62,  file=vsbrrs(j))
      open(unit=71,  file=vsbirm(i))
      open(unit=72,  file=vsbirm(j))
      open(unit=81,  file=vsbirf(i))
      open(unit=82,  file=vsbirf(j))
      call ibir(51, 52, 61, 62, 71, 72, 81, 82)
```

```
            if(ierror.ne.0) then
              routin = 'ibir'
              goto 9999
              endif
            close(51)
            close(52)
            close(61)
            close(62)
            close(71)
            close(72)
            close(81)
            close(82)
c
            open(unit=91,  file=vsdthm(i))
            open(unit=92,  file=vsdthm(j))
            call idth(91, 92, ndthm, dwidm0, dwdsm1, dthwdm, dthrim, dthrsm)
            if(ierror.ne.0) then
              routin = 'idth males'
              goto 9999
              endif
            close(91)
            close(92)
c
            open(unit=101, file=vsdthf(i))
            open(unit=102, file=vsdthf(j))
            call idth(101, 102, ndthf, dwidf0, dwdsf1, dthwdf, dthrif, dthrsf)
            if(ierror.ne.0) then
              routin = 'idth females'
              goto 9999
              endif
            close(101)
            close(102)
c
            open(unit=111, file=vsnmar(i))
            open(unit=112, file=vsnmar(j))
            call imar(111, 112)
            if(ierror.ne.0) then
              routin = 'imar'
              goto 9999
              endif
            close(111)
            close(112)
c
c  call the two-sex population simulator
c
            if(extrap.eq.0.0) time2 = time1
```

```
                  call twosex(time1, time2, deltat, nstep, nprint, ierror)
                  if(ierror.ne.0) then
                    routin = 'twosex'
                    goto 9999
                    endif
c
                  stop
c
c  error output
c
   9999           write(6,9998) ierror, routin
   9998           format('error number', i3, ' in ',a)
                  stop
                  end
c
c===========================================
c
                  subroutine iu0(nunit1, nunit2, n, width, u01, u02)
c
c  initialize the distribution of males or females
c
c------------------------------------------------------------------
                  common/init/ time1, time2, extrap, maxdim, ierror
                  dimension u01(0:25), u02(0:25)
                  common/temp3/ input(0:25,0:25)
                  integer wid1, wid2, tim1, tim2, time1, time2
c
c  make sure files have compatible data; else ierror = 1 or 2
c
                  read(nunit1,100) n,  wid1, tim1
                  read(nunit2,100) n2, wid2, tim2
   100            format(i16,/,i16,/,i16)
                  if((n.ne.n2).or.(wid1.ne.wid2)) ierror = 1
                  if((tim1.ne.time1).or.(tim2.ne.time2)) ierror = 2
                  if(n.gt.maxdim) ierror = 9
                  if(ierror.ne.0) return
                  n = n - 1
                  width = float(wid1)
c
c  read in data and adjust the raw data
c
                  read(nunit1,110) (input(k,0),k=0,n)
                  do 200 k=0,n
   200            u01(k) = float(input(k,0)) / width
                  read(nunit2,110) (input(k,0),k=0,n)
                  do 210 k=0,n
```

```
 210        u02(k) = float(input(k,0)) / width
 110        format(i16)
c
            return
            end
c
c=========================================
c
            subroutine ic0(nunit1, nunit2, n, width, c01, c02)
c
c  initialize the couples distribution
c
c--------------------------------------------------------------
            common/init/ time1, time2, extrap, maxdim, ierror
            dimension c01(0:25,0:25), c02(0:25,0:25)
            common/temp3/ input(0:25,0:25)
            integer wid1, wid2, tim1, tim2, time1, time2
c
c  make sure files have compatible data; else ierror = 1 or 2
c
            read(nunit1,100) n,  wid1, tim1
            read(nunit2,100) n2, wid2, tim2
 100        format(i16,/,i16,/,i16)
            if((n.ne.n2).or.(wid1.ne.wid2)) ierror = 1
            if((tim1.ne.time1).or.(tim2.ne.time2)) ierror = 2
            if(n.gt.maxdim) ierror = 9
            if(ierror.ne.0) return
            n = n - 1
            width = float(wid1)
c
c  read in data and adjust the raw data
c
            widsq = width*width
            read(nunit1,110) ((input(i,j),i=0,n),j=0,n)
            do 200 j=0,n
            do 200 i=0,n
 200        c01(i,j) = float(input(i,j)) / widsq
            read(nunit2,110) ((input(i,j),i=0,n),j=0,n)
            sum = 0.0
            do 101 iii = 0,n
                do 102 jjj = 0,n
                    sum = sum + input(iii,jjj)
 102        continue
 101        continue
            print *, 'SUM of 1980 couples:'
            print '(f20.2)', sum
```

```
                do 210 j=0,n
                do 210 i=0,n
  210           c02(i,j) = float(input(i,j)) / widsq
  110           format(i16)
c
                return
                end
c
c=======================================
c
                subroutine iann(nunit1, nunit2, nunit3, nunit4)
c
c  initialize the annulment/divorce data
c
c------------------------------------------------------------
                common/init/ time1, time2, extrap, maxdim, ierror
                common/anncom/ n, width, annri(0:25,0:25), annrs(0:25,0:25)
                common/temp1/ a1(0:25), a2(0:25)
                common/temp2/ a3(0:25), a4(0:25)
                common/mincom/ annmin, dthmin
                common/rndcom/ zero
                common/temp3/ input(0:25,0:25)
                integer wid1, wid2, wid3, wid4,
          $        tim1, tim2, tim3, tim4, time1, time2
c
c  make sure files have compatible data; else ierror = 1 or 2
c
                read(nunit1,100) n,  wid1, tim1
                read(nunit2,100) n2, wid2, tim2
                read(nunit3,100) n3, wid3, tim3
                read(nunit4,100) n4, wid4, tim4
  100           format(i16,/,i16,/,i16)
                if((n.ne.n2).or.(n.ne.n3).or.(n.ne.n4).or.
          $     (wid1.ne.wid2).or.(wid1.ne.wid3).or.
          $                      (wid1.ne.wid4)) ierror = 1
                if((tim1.ne.time1).or.(tim3.ne.time1).or.
          $     (tim2.ne.time2).or.(tim4.ne.time2)) ierror = 2
                if(n.gt.maxdim) ierror = 9
                if(ierror.ne.0) return
                n = n - 1
                width = float(wid1)
c
c  read in data
c  note that initial zeros are needed in the data
c
                read(nunit1,110) itot, (input(i,0),i=0,n)
```

```
      tot1 = float(itot)
      do 200 i=0,n
200   a1(i) = float(input(i,0))
      read(nunit2,110) itot, (input(i,0),i=0,n)
      tot2 = float(itot)
      do 210 i=0,n
210   a2(i) = float(input(i,0))
      read(nunit3,110) itot, (input(j,0),j=0,n)
      do 220 j=0,n
220   a3(j) = float(input(j,0))
      read(nunit4,110) itot, (input(j,0),j=0,n)
      do 230 j=0,n
230   a4(j) = float(input(j,0))
110   format(i16)
c
c determine totals of known cases
c
      totkn1 = 0.0
      totkn2 = 0.0
      totkn3 = 0.0
      totkn4 = 0.0
      do 10 k=0,n
        totkn1 = totkn1 + a1(k)
        totkn2 = totkn2 + a2(k)
        totkn3 = totkn3 + a3(k)
10    totkn4 = totkn4 + a4(k)
c
c form product rates
c
      klmax  = ifix(width - 1.0 + zero)
      x      = zero
      do 20 i = 0,n
        y = zero
        do 30 j = 0,n
          totc0  = 0.0
          totc0p = 0.0
          xx     = x
          do 40 k = 0,klmax
            yy = y
            do 50 l = 0,klmax
              totc0  = totc0  + c0(xx,yy)
              totc0p = totc0p + c0p(xx,yy)
              yy = yy + 1.0
50    continue
      xx = xx + 1.0
40    continue
```

```
                  denom = totkn1 * totkn3 * totc0
                  if(denom.eq.0.0) then
                     annri(i,j) = annmin
                    else
                     annri(i,j) = a1(i) * a3(j) * tot1 / denom
                     if(annri(i,j).lt.annmin) annri(i,j) = annmin
                    endif
                  denom = totkn2 * totkn4 * totc0p
                  if(denom.eq.0.0) then
                     annrs(i,j) = extrap * (annri(i,j) - annmin)
                    else
                     annrs(i,j) = a2(i) * a4(j) * tot2 / denom
                     if(annrs(i,j).lt.annmin) annrs(i,j) = annmin
                     annrs(i,j) = extrap * (annri(i,j) - annrs(i,j))
                     avers      = avers + annrs(i,j)
                     navers     = navers + 1
                    endif
                  y = y + width
   30             continue
                  x = x + width
   20             continue
c
                  return
                  end
c
c=========================================
c
                  subroutine ibir (nunit1, nunit2, nunit3, nunit4,
              $                        nunit5, nunit6, nunit7, nunit8)
c
c  initialize married (coupled) and unmarried (single) birth rates
c
c--------------------------------------------------------------------
                  common/init/ time1, time2, extrap, maxdim, ierror
                  common/bircom/ n, width
                  common/brcmcm/ brcmri(0:25,0:25), brcmrs(0:25,0:25)
                  common/brcfcm/ brcfri(0:25,0:25), brcfrs(0:25,0:25)
                  common/brsmcm/ brsmri(0:25), brsmrs(0:25)
                  common/brsfcm/ brsfri(0:25), brsfrs(0:25)
                  common/temp1/ brci(0:25), brcs(0:25)
                  common/rndcom/ zero
                  common/temp3/ input(0:25,0:25)
                  integer wid1, wid2, wid3, wid4, wid5, wid6, wid7, wid8,
              $      tim1, tim2, tim3, tim4, tim5, tim6, tim7, tim8,
              $      time1, time2
c
```

```
c  make sure files have compatible data; else ierror = 1 or 2
c
              read(nunit1,1000) n,  wid1, tim1
              read(nunit2,1000) n2, wid2, tim2
              read(nunit3,1000) n3, wid3, tim3
              read(nunit4,1000) n4, wid4, tim4
              read(nunit5,1000) n5, wid5, tim5
              read(nunit6,1000) n6, wid6, tim6
              read(nunit7,1000) n7, wid7, tim7
              read(nunit8,1000) n8, wid8, tim8
    1000      format(i16,/,i16,/,i16)
              if((n.ne.n2).or.(n.ne.n3).or.(n.ne.n4).or.(n.ne.n5).or.
           $  (n.ne.n6).or.(n.ne.n7).or.(n.ne.n8)) ierror = 1
              if((wid1.ne.wid2).or.(wid1.ne.wid3).or.
           $  (wid1.ne.wid4).or.(wid1.ne.wid5).or.
           $  (wid1.ne.wid6).or.(wid1.ne.wid7).or.
           $  (wid1.ne.wid8)) ierror = 1
              if((tim1.ne.time1).or.(tim2.ne.time2).or.
           $  (tim3.ne.time1).or.(tim4.ne.time2).or.
           $  (tim5.ne.time1).or.(tim6.ne.time2).or.
           $  (tim7.ne.time1).or.(tim8.ne.time2)) ierror = 2
              if(n.gt.maxdim) ierror = 9
              if(ierror.ne.0) return
              n = n - 1
              width = float(wid1)
c
c  read in data
c
              read(nunit1,1100) (brci(j),j=0,n)
              read(nunit2,1100) (brcs(j),j=0,n)
              read(nunit3,1100) (brsmri(j),j=0,n)
              read(nunit4,1100) (brsmrs(j),j=0,n)
              read(nunit5,1200) ((input(i,j),i=0,n),j=0,n)
              do 210 j=0,n
              do 210 i=0,n
    210       brcmri(i,j) = float(input(i,j))
              read(nunit6,1200) ((input(i,j),i=0,n),j=0,n)
              do 220 j=0,n
              do 220 i=0,n
    220       brcmrs(i,j) = float(input(i,j))
              read(nunit7,1200) ((input(i,j),i=0,n),j=0,n)
              do 230 j=0,n
              do 230 i=0,n
    230       brcfri(i,j) = float(input(i,j))
              read(nunit8,1200) ((input(i,j),i=0,n),j=0,n)
              do 240 j=0,n
```

```
          do 240 i=0,n
  240     brcfrs(i,j) = float(input(i,j))
  1100    format(g16.8)
  1200    format(i16)
c
c adjust the raw data
c
          klmax  = ifix(width - 1.0 + zero)
          y      = zero
          do 20 j = 0,n
            tc0j   = 0.0
            tc0jp  = 0.0
            tbrmj  = 0.0
            tbrmjp = 0.0
            tbrfj  = 0.0
            tbrfjp = 0.0
            x      = zero
c total number of couples
          do 30 i = 0,n
            tc0ij  = 0.0
            tc0ijp = 0.0
            xx     = x
            do 40 k = 0,klmax
              yy = y
              do 50 l = 0,klmax
                tc0ij  = tc0ij  + c0(xx,yy)
                tc0ijp = tc0ijp + c0p(xx,yy)
                yy = yy + 1.0
  50      continue
          xx = xx + 1.0
  40      continue
c
          tc0j   = tc0j   + tc0ij
          tc0jp  = tc0jp  + tc0ijp
          tbrmj  = tbrmj  + brcmri(i,j)
          tbrmjp = tbrmjp + brcmrs(i,j)
          tbrfj  = tbrfj  + brcfri(i,j)
          tbrfjp = tbrfjp + brcfrs(i,j)
c
          if(tc0ij.eq.0.0) then
              brcmri(i,j) = 0.0
              brcfri(i,j) = 0.0
            else
              brcmri(i,j) = brcmri(i,j) * brci(j) / tc0ij
              brcfri(i,j) = brcfri(i,j) * brci(j) / tc0ij
            endif
```

```
          if(tc0ijp.eq.0.0) then
              brcmrs(i,j) = 0.0
              brcfrs(i,j) = 0.0
            else
              brcmrs(i,j) = brcmrs(i,j) * brcs(j) / tc0ijp
              brcfrs(i,j) = brcfrs(i,j) * brcs(j) / tc0ijp
            endif
          x = x + width
30        continue
          tbirj = tbrmj + tbrfj
          if(tbirj.eq.0.0) then
              do 60 i = 0,n
                brcmri(i,j) = 0.0
60        brcfri(i,j) = 0.0
              brsmri(j) = 0.0
              brsfri(j) = 0.0
            else
              do 70 i = 0,n
                brcmri(i,j) = brcmri(i,j) * tc0j / tbirj
70        brcfri(i,j) = brcfri(i,j) * tc0j / tbirj
              brsfri(j) = brsmri(j) * tbrfj / tbirj
              brsmri(j) = brsmri(j) - brsfri(j)
            endif
          tbirjp = tbrmjp + tbrfjp
          if(tbirjp.eq.0.0) then
              do 80 i = 0,n
                brcmrs(i,j) = 0.0
80        brcfrs(i,j) = 0.0
              brsmrs(j) = 0.0
              brsfrs(j) = 0.0
            else
              do 90 i = 0,n
                brcmrs(i,j) = brcmrs(i,j) * tc0jp / tbirjp
90        brcfrs(i,j) = brcfrs(i,j) * tc0jp / tbirjp
              brsfrs(j) = brsmrs(j) * tbrfjp / tbirjp
              brsmrs(j) = brsmrs(j) - brsfrs(j)
            endif
          y = y + width
20        continue
c
          do 100 j = 0,n
            brsmrs(j) = extrap * (brsmri(j) - brsmrs(j))
            brsfrs(j) = extrap * (brsfri(j) - brsfrs(j))
            do 100 i = 0,n
              brcmrs(i,j) = extrap * (brcmri(i,j) - brcmrs(i,j))
              brcfrs(i,j) = extrap * (brcfri(i,j) - brcfrs(i,j))
```

```
      100           continue
c
                    return
                    end
c
c==========================================
c
                    subroutine idth (nunit1, nunit2,
               $                 n, width0, wdsum1, width, dthri, dthrs)
c
c  initialize death rates
c
c----------------------------------------------------------------------------
                    common/init/ time1, time2, extrap, maxdim, ierror
                    dimension dthri(0:25), dthrs(0:25)
                    integer wid01, wids11, wid1, wid02, wids12, wid2,
               $       tim1, tim2, time1, time2
c
c  make sure files have compatible data; else ierror = 1 or 2
c
                    read(nunit1,100) n,  wid01, wids11, wid1, tim1
                    read(nunit2,100) n2, wid02, wids12, wid2, tim2
      100           format(i16,4(/,i16))
                    if((n.ne.n2).or.(wid01.ne.wid02).or.
               $   (wids11.ne.wids12).or.(wid1.ne.wid2)) ierror = 1
                    if((tim1.ne.time1).or.(tim2.ne.time2)) ierror = 2
                    if(n.gt.maxdim) ierror = 9
                    if(ierror.ne.0) return
                    n = n - 1
                    width0 = float(wid01)
                    wdsum1 = float(wids11)
                    width  = float(wid1)
c
c  read in data
c
                    read(nunit1,110) (dthri(k),k=0,n)
                    read(nunit2,110) (dthrs(k),k=0,n)
      110           format(g16.8)
c
c  adjust the raw data
c
                    do 10 k=0,n
      10            dthrs(k) = extrap * (dthri(k) - dthrs(k))
c
                    return
                    end
```

```
c
c==========================================
c
                    subroutine imar (nunit1, nunit2)
c
c  initialize preference functions g(a,b) and h(a,b)
c
c------------------------------------------------------------------
                    common/init/ time1, time2, extrap, maxdim, ierror
                    common/marcom/ n, width, nmari(0:25,0:25), nmars(0:25,0:25)
                    common/funcg/ gi(0:25,0:25), gs(0:25,0:25)
                    common/funch/ hi(0:25,0:25), hs(0:25,0:25)
                    common/temp1/ rm1(0:25), rf1(0:25)
                    common/temp2/ rm2(0:25), rf2(0:25)
                    common/rndcom/ zero
                    integer wid1, wid2, wid3, wid4,
               $       tim1, tim2, tim3, tim4, time1, time2
c
c  make sure files have compatible data; else ierror = 1 or 2
c
                    read(nunit1,100) n,  wid1,  tim1
                    read(nunit2,100) n2, wid2, tim2
    100             format(i16,/,i16,/,i16)
                    if((n.ne.n2).or.(wid1.ne.wid2)) ierror = 1
                    if((tim1.ne.time1)) ierror = 2
                    if(n.gt.maxdim) ierror = 9
                    if(ierror.ne.0) return
                    n = n - 1
                    width = float(wid1)
c
c  read in data
c
                    read(nunit1,110) ((nmari(i,j),j=0,n),i=0,n)
                    read(nunit2,110) ((nmars(i,j),j=0,n),i=0,n)
    110             format(i16)
c
c  calculate total number of single males and of single females
                    lmax   = ifix(width - 1.0 + zero)
                    mmax   = (n + 1) * (lmax + 1) - 1
                    totmm  = 0.0
                    totmmp = 0.0
                    totmf  = 0.0
                    totmfp = 0.0
                    z      = zero
                    do 20 k = 0,n
                      tum0i  = 0.0
```

```
            tum0pi = 0.0
            tuf0j  = 0.0
            tuf0pj = 0.0
            tc0i   = 0.0
            tc0pi  = 0.0
            tc0j   = 0.0
            tc0pj  = 0.0
            zz     = z
            do 30 l = 0,lmax
              ww = zero
              do 40 m = 0,mmax
                tc0i  = tc0i  + c0(zz,ww)
                tc0pi = tc0pi + c0p(zz,ww)
                tc0j  = tc0j  + c0(ww,zz)
                tc0pj = tc0pj + c0p(ww,zz)
                ww = ww + 1.0
 40           continue
              tum0i  = tum0i  + um0(zz)
              tum0pi = tum0pi + um0p(zz)
              tuf0j  = tuf0j  + uf0(zz)
              tuf0pj = tuf0pj + uf0p(zz)
              zz     = zz + 1.0
 30         continue
            smi  = tum0i  - tc0i
            smpi = tum0pi - tc0pi
            sfj  = tuf0j  - tc0j
            sfpj = tuf0pj - tc0pj
            if(smi .lt.0.0) smi  = 0.0
            if(smpi.lt.0.0) smpi = 0.0
            if(sfj .lt.0.0) sfj  = 0.0
            if(sfpj.lt.0.0) sfpj = 0.0
            rm1(k) = smi
            rm2(k) = smpi
            rf1(k) = sfj
            rf2(k) = sfpj
            z = z + width
 20       continue
c
c  compute preference functions g(a,b,t) and h(a,b,t)
c
                  do 55 i=0,n
                     do 65 j=0,n
          if(rf1(i).lt.1.e-7) then
          gi(i,j) = 0.
          else
                           gi(i,j) = float(nmari(i,j))/(width*rf1(i))
```

```
      endif
      if(rf2(i).lt.1.e-7) then
      gs(i,j) = 0.
      else
      gs(i,j) = (float(nmars(i,j))/(width*rf2(i)) -
     *          gi(i,j))/abs(tim1-tim2)
      endif
      if(rm1(j).lt.1.e-7) then
      hi(i,j) = 0.
      else
      hi(i,j) = float(nmari(i,j))/(width*rm1(j))
      endif
      if(rm2(j).lt.1.e-7) then
      hs(i,j) = 0.
      else
      hs(i,j) = (float(nmars(i,j))/(width*rm2(j)) -
     *          hi(i,j))/abs(tim1-tim2)
      endif
65    continue

         return
         end
c
c==========================================
c
      function um0(x)
      common/um0com/ n, width, um01(0:25), um02(0:25)
      i = ifix(x / width)
      if(i.gt.n) then
          um0 = 0.0
        else
          um0 = um01(i)
        endif
      return
      end
c
c==========================================
c
      function um0p(x)
      common/um0com/ n, width, um01(0:25), um02(0:25)
      i = ifix(x / width)
      if(i.gt.n) then
          um0p = 0.0
        else
          um0p = um02(i)
        endif
```

```
      return
      end
c
c==========================================
c
      function uf0(y)
      common/uf0com/ n, width, uf01(0:25), uf02(0:25)
      j = ifix(y / width)
      if(j.gt.n) then
          uf0 = 0.
        else
          uf0 = uf01(j)
        endif
      return
      end
c
c==========================================
c
      function uf0p(y)
      common/uf0com/ n, width, uf01(0:25), uf02(0:25)
      j = ifix(y / width)
      if(j.gt.n) then
          uf0p = 0.0
        else
          uf0p = uf02(j)
        endif
      return
      end
c
c==========================================
c
      function c0(x,y)
      common/c0com/ n, width, c01(0:25,0:25), c02(0:25,0:25)
      i = ifix(x / width)
      j = ifix(y / width)
      if((i.gt.n).or.(j.gt.n)) then
          c0 = 0.0
        else
          c0 = c01(i,j)
        endif
      return
      end
c
c==========================================
c
      function c0p(x,y)
```

```
      common/c0com/ n, width, c01(0:25,0:25), c02(0:25,0:25)
      i = ifix(x / width)
      j = ifix(y / width)
      if((i.gt.n).or.(j.gt.n)) then
         c0p = 0.0
        else
         c0p = c02(i,j)
        endif
      return
      end
c
c=========================================
c
      function betam(x,y,t)
      common/bircom/ n, width
      common/brcmcm/ brcmri(0:25,0:25), brcmrs(0:25,0:25)
      i     = min(n, ifix(x / width))
      j     = min(n, ifix(y / width))
      betam = brcmri(i,j) + t * brcmrs(i,j)
      if(betam.lt.0.0) betam = 0.0
      if(betam.gt.1.0) betam = 1.0
      return
      end
c
c=========================================
c
      function betaf(x,y,t)
      common/bircom/ n, width
      common/brcfcm/ brcfri(0:25,0:25), brcfrs(0:25,0:25)
      i     = min(n, ifix(x / width))
      j     = min(n, ifix(y / width))
      betaf = brcfri(i,j) + t * brcfrs(i,j)
      if(betaf.lt.0.0) betaf = 0.0
      if(betaf.gt.1.0) betaf = 1.0
      return
      end
c
c=========================================
c
      function betasm(y,t)
      common/bircom/ n, width
      common/brsmcm/ brsrim(0:25), brsrsm(0:25)
      j      = min(n, ifix(y / width))
      betasm = brsrim(j) + t * brsrsm(j)
      if(betasm.lt.0.0) betasm = 0.0
      if(betasm.gt.1.0) betasm = 1.0
```

```
                  return
                  end
c
c===========================================
c
                  function betasf(y,t)
                  common/bircom/ n, width
                  common/brsfcm/ brsrif(0:25), brsrsf(0:25)
                  j      = min(n, ifix(y / width))
                  betasf = brsrif(j) + t * brsrsf(j)
                  if(betasf.lt.0.0) betasf = 0.0
                  if(betasf.gt.1.0) betasf = 1.0
                  return
                  end
c
c===========================================
c
                  function delm(x,t)
                  common/mincom/ annmin, dthmin
                  common/dthmcm/ n, width0, widsm1, width,
               $            dthrim(0:25), dthrsm(0:25)
                  if(x.lt.width0) then
                      i = 0
                    else
                      if(x.lt.widsm1) then
                        i = 1
                      else
                        i = min(n, (ifix((x - widsm1) / width) + 2))
                      endif
                    endif
                  delm = dthrim(i) + t * dthrsm(i)
                  if(delm.lt.dthmin) delm = dthmin
                  if(delm.gt.1.0) delm = 1.0
                  return
                  end
c
c===========================================
c
                  function delf(y,t)
                  common/mincom/ annmin, dthmin
                  common/dthfcm/ n, width0, widsm1, width,
               $            dthrif(0:25), dthrsf(0:25)
                  if(y.lt.width0) then
                      j = 0
                    else
                      if(y.lt.widsm1) then
```

```
                j = 1
              else
                j = min(n, (ifix((y - widsm1) / width) + 2))
              endif
            endif
          delf = dthrif(j) + t * dthrsf(j)
          if(delf.lt.dthmin) delf = dthmin
          if(delf.gt.1.0) delf = 1.0
          return
          end
c
c=========================================
c
          real function mu(x,y,t,sm,sf)
          common/marcom/ n, width, nmari(0:25,0:25), nmars(0:25,0:25)
          common/funcg/ gi(0:25,0:25), gs(0:25,0:25)
          common/funch/ hi(0:25,0:25), hs(0:25,0:25)
          real g,h
c
          if((sm.le.0.0).or.(sf.le.0.0)) then
            mu = 0.0
            return
            endif
c
c  evaluate the real functions g(x,y,t) and h(x,y,t)
c
          i  = min(n, ifix(x / width))
          j  = min(n, ifix(y / width))

          g = gi(i,j) + t * gs(i,j)
          h = hi(i,j) + t * hs(i,j)

          if (g.lt.0.0) g = 0.0
          if (g.gt.1.0) g = 1.0
          if (h.lt.0.0) h = 0.0
          if (h.gt.1.0) h = 1.0

          if (abs(h * sm + g * sf).lt.(1.E-08)) then
            mu = 0.0
          else
            mu = 2.0 * h * sm * g* sf / (h * sm + g * sf)
          endif
          if(mu.gt.sm) mu = sm
          if(mu.gt.sf) mu = sf
c
          return
```

```
                    end
c
c==============================================
c     weighted harmonic mean with preferences

                    real function mu1(x,y,t,sm,sf)
                    common/marcom/ n, width, nmari(0:25,0:25), nmars(0:25,0:25)
                    common/funcg/ gi(0:25,0:25), gs(0:25,0:25)
                    common/funch/ hi(0:25,0:25), hs(0:25,0:25)
                    real g,h
c
                    if((sm.le.0.0).or.(sf.le.0.0)) then
                      mu1 = 0.0
                      return
                      endif
c
c  evaluate the real functions g(x,y,t) and h(x,y,t)
c
                    i  = min(n, ifix(x / width))
                    j  = min(n, ifix(y / width))

                    g = gi(i,j) + t * gs(i,j)
                    h = hi(i,j) + t * hs(i,j)

                    if (g.lt.0.0) g = 0.0
                    if (g.gt.1.0) g = 1.0
                    if (h.lt.0.0) h = 0.0
                    if (h.gt.1.0) h = 1.0
c
                    if (abs(0.5 * h * sm + 0.5 * g * sf).lt.1E-08) then
                      mu1 = 0.0
                    else
                      mu1 = 1.0 * h * sm * g* sf / (0.5 * h * sm + 0.5 * g * sf)
                    endif
                    if(mu1.gt.sm) mu1 = sm
                    if(mu1.gt.sf) mu1 = sf
c
                    return
                    end
c
c==============================================
c     weighted geometric mean with preferences

                    real function mu2(x,y,t,sm,sf)
                    common/marcom/ n, width, nmari(0:25,0:25), nmars(0:25,0:25)
                    common/funcg/ gi(0:25,0:25), gs(0:25,0:25)
```

```
                common/funch/ hi(0:25,0:25), hs(0:25,0:25)
                real g,h
c
                if ((sm.le.0.0).or.(sf.le.0.0)) then
                  mu2 = 0.0
                  return
                  endif
c
c  evaluate the real functions g(x,y,t) and h(x,y,t)
c
                i  = min(n, ifix(x / width))
                j  = min(n, ifix(y / width))

                g = gi(i,j) + t * gs(i,j)
                h = hi(i,j) + t * hs(i,j)

                if (g.lt.0.0) g = 0.0
                if (g.gt.1.0) g = 1.0
                if (h.lt.0.0) h = 0.0
                if (h.gt.1.0) h = 1.0
c
                if (((h * sm).ge.0.0).and.((g * sf).ge.0.0)) then
                  mu2 = ((h * sm)**0.4) * ((g* sf)**0.6)
                else
                  print*,'error in mu2'
                endif
                if(mu2.gt.sm) mu2 = sm
                if(mu2.gt.sf) mu2 = sf
c
                return
                end
c
c=========================================
c     the min function with preferences

                real function mu3(x,y,t,sm,sf)
                common/marcom/ n, width, nmari(0:25,0:25), nmars(0:25,0:25)
                common/funcg/ gi(0:25,0:25), gs(0:25,0:25)
                common/funch/ hi(0:25,0:25), hs(0:25,0:25)
                real g,h
c
                if((sm.le.0.0).or.(sf.le.0.0)) then
                  mu3 = 0.0
                  return
                  endif
c
```

```
c  evaluate the real functions g(x,y,t) and h(x,y,t)
c
              i  = min(n, ifix(x / width))
              j  = min(n, ifix(y / width))

              g = gi(i,j) + t * gs(i,j)
              h = hi(i,j) + t * hs(i,j)

              if (g.lt.0.0) g = 0.0
              if (g.gt.1.0) g = 1.0
              if (h.lt.0.0) h = 0.0
              if (h.gt.1.0) h = 1.0
c
              if (((h * sm).lt.0.0).or.((g * sf).lt.0.0)) then
                print*,'error in mu3'
              else
                if ((h * sm).le.(g * sf)) then
                  mu3 = h * sm
                else
                  mu3 = g * sf
                endif
              endif
              if(mu3.gt.sm) mu3 = sm
              if(mu3.gt.sf) mu3 = sf
c
              return
              end
c
c=========================================
c

c    geometric mean with preferences

              real function mu4(x,y,t,sm,sf)
              common/marcom/ n, width, nmari(0:25,0:25), nmars(0:25,0:25)
              common/funcg/ gi(0:25,0:25), gs(0:25,0:25)
              common/funch/ hi(0:25,0:25), hs(0:25,0:25)
              real g,h
c
              if ((sm.le.0.0).or.(sf.le.0.0)) then
                mu4 = 0.0
                return
                endif
c
c  evaluate the real functions g(x,y,t) and h(x,y,t)
c
```

```
      i  = min(n, ifix(x / width))
      j  = min(n, ifix(y / width))

      g = gi(i,j) + t * gs(i,j)
      h = hi(i,j) + t * hs(i,j)

      if (g.lt.0.0) g = 0.0
      if (g.gt.1.0) g = 1.0
      if (h.lt.0.0) h = 0.0
      if (h.gt.1.0) h = 1.0
c
      if (((h * sm).ge.0.0).and.((g * sf).ge.0.0)) then
        mu4 = ((h * sm)**0.5) * ((g* sf)**0.5)
      else
        print*,'error in mu4'
      endif
      if(mu4.gt.sm) mu4 = sm
      if(mu4.gt.sf) mu4 = sf

      return
      end
c
c=========================================
c
      function sigma(x,y,t)

      common/anncom/ n, width, annri(0:25,0:25), annrs(0:25,0:25)
      common/mincom/ annmin, dthmin

c
c  evaluate the function alpha(x,y,t)
c
      i     = min(n, ifix(x / width))
      j     = min(n, ifix(y / width))
      alpha = annri(i,j) + t * annrs(i,j)
      if(alpha.lt.0.0) alpha = annmin
      if(alpha.gt.1.0) alpha = 1.0
c
      dm    = delm(x,t)
      df    = delf(y,t)
      sigma = dm + df - dm * df + alpha
      if(sigma.gt.1.0) sigma = 1.0
c
      return
      end
c
c=========================================
```

```
c
                  subroutine twosex(time1, time2, deltat, nstep, nprint, ierror)
c
c=========================================
c
                  common/um0com/ num0, um0wid, um01(0:25), um02(0:25)
                  common/uf0com/ nuf0, uf0wid, uf01(0:25), uf02(0:25)
                  common/c0com/ nc0, c0wid, c01(0:25,0:25), c02(0:25,0:25)
                  common/soln/ um(0:1880), uf(0:1880), c(0:1880,0:1880),
              $          sm(0:1880), sf(0:1880), p, msupp, n, t
                  integer time1, time2, iiii
                  real mu, mu1, mu2, mu3, mu4
                  data mxmsup /1880/
c
                  tau = nstep * deltat
                  if((nstep.lt.0).or.(deltat.le.0.0)) then
                    ierror = 11
                    return
                    endif
c
c  initial output
c
                  write(6,1000) tau, nstep, deltat, time1, time2
       1000       format('max time (scaled from 0) =',f9.4,5x,/,
              $     'number of time steps =',i4,5x,'delta t =',f9.6,5x,/,
              $     'starting year =',i5,5x,'extrap/interp year =',i5)
                  write(6,1010)
       1010       format ('------------------------------',
              $     '------------------------------')
c
c=========================================
c  initialization
c------------------------------------------------------------------------
c
c  initialize the sex and couples distributions
c
                  n     = 0
                  t     = 0.0
                  msupp = 0
                  do 10 k = 0,mxmsup
                    z      = k * deltat
                    um(k) = um0(z)
                    uf(k) = uf0(z)
                    if((um(k).ne.0.0).or.(uf(k).ne.0.0)) then
                      if(k.gt.msupp) msupp = k
                      endif
```

```
                do 20 i = 0,mxmsup
                  x      = i * deltat
                  c(i,k) = c0(x,z)
                  if(c(i,k).ne.0.0) then
                    if(i.gt.msupp) msupp = i
                    if(k.gt.msupp) msupp = k
                    endif
   20           continue
   10           continue
                if((msupp + nstep).gt.mxmsup) then
                  ierror = 19
                  return
                  endif
c
c  initialize the singles distributions and the total population
c
                p   = 0.0
                do 30 k = 0,msupp
                  cmk = 0.0
                  cfk = 0.0
                  p = p + um(k) + uf(k)
                  do 40 l = 0,msupp
                    cmk = cmk + c(k,l)
                    cfk = cfk + c(l,k)
   40           continue
                  sm(k) = um(k) - cmk * deltat
                  sf(k) = uf(k) - cfk * deltat
                  if(sm(k).lt.0.0) sm(k) = 0.0
                  if(sf(k).lt.0.0) sf(k) = 0.0
   30           continue
                p = p * deltat
c
c  call for output of initial values
c
                call outp(deltat)
c
c========================================
c  loop on time-steps:
c------------------------------------------------------------
c
                freq   = float(nstep) / float(nprint)
                prstep = freq
                nextpr = ifix(prstep + 0.5)
                do 900 n = 1,nstep
                  t      = t + deltat
                  msupp = msupp + 1
```

```
c
c-----------------------------------------------------------------
c  solve for the couples distribution
c.................................................................
c
             do 110 j = msupp,1,-1
               sfp = sf(j-1)
               y   = j * deltat

               do 110 i = msupp,1,-1
                 smp   = sm(i-1)
                 x     = i * deltat
                 c(i,j) = (c(i-1,j-1) + mu2(x,y,t,smp,sfp) * deltat) /
     $                           (1.0 + sigma(x,y,t) * deltat)
  110        continue
c
             do 120 k=0,msupp
               c(k,0)=0.0
               c(0,k)=0.0
  120        continue
c
c-----------------------------------------------------------------
c  solve for the distributions of males, females, and singles
c.................................................................
c
c  deaths
c
             do 210 k = msupp,1,-1
               z     = k * deltat
               um(k) = um(k-1) / (1.0 + delm(z,t) * deltat)
               uf(k) = uf(k-1) / (1.0 + delf(z,t) * deltat)
  210        continue
c
c  singles (and total population)
c
                     p    = 0.0
             guys = 0.0
             gals = 0.0
             boys = 0.0
             girls = 0.0
             do 242 i=0,ifix(1./deltat)/2-1
               boys = boys+um(i)+4.*um(i+1)+um(i+2)
               girls = girls+uf(i)+4.*uf(i+1)+uf(i+2)
  242        continue
             boys = boys*deltat/3.
             girls = girls*deltat/3.
```

```
                 do 230 k = 1,msupp
                   cmk = 0.0
                   cfk = 0.0
                   p = p + um(k) + uf(k)
            guys = guys + um(k)
            gals = gals + uf(k)
                   do 240 l = 1,msupp
                     cmk = cmk + c(k,l)
                     cfk = cfk + c(l,k)
240         continue
            sm(k) = um(k) - cmk * deltat
            sf(k) = uf(k) - cfk * deltat
            if(sm(k).lt.0.0) sm(k) = 0.0
            if(sf(k).lt.0.0) sf(k) = 0.0
230         continue
c
c  births
c
            bms = 0.0
            bmc = 0.0
            bfs = 0.0
            bfc = 0.0
            do 220 j = 1,msupp
              y   = j * deltat
              bms = bms + betasm(y,t) * sf(j)
              bfs = bfs + betasf(y,t) * sf(j)
              do 320 i = 1,msupp
                x   = i * deltat
                bmc = bmc + betam(x,y,t) * c(i,j)
                bfc = bfc + betaf(x,y,t) * c(i,j)
320         continue
220         continue
                 um(0) = (bmc * deltat + bms) * deltat
                 uf(0) = (bfc * deltat + bfs) * deltat
                 sm(0) = um(0)
                 sf(0) = uf(0)
                 p = (p + um(0) + uf(0)) * deltat
            guys = (guys + um(0)) * deltat
            gals = (gals + uf(0)) * deltat
c write(6,*)p

c
c------------------------------------------------------------------
c  call the output routine
c..................................................................
c
```

```
                if(n.eq.nextpr) then
                  call outp(deltat)
                  prstep = prstep + freq
                  nextpr = ifix(prstep + 0.5)
                  endif
c
c----------------------------------------------------------------------
c
  900           continue
c
c=========================================
c
                print *, 'TOTAL MALES:'
                print '(f20.2)', guys
                print *, 'TOTAL FEMALES:'
                print '(f20.2)', gals
                print *, 'MALES under 1 YEAR of AGE:'
                print '(f20.2)', boys
                print *, 'FEMALES under 1 YEAR of AGE:'
                print '(f20.2)', girls
                print *, 'NEWBORN MALES:'
                print '(f20.2)', um(0)
                print *, 'NEWBORN FEMALES:'
                print '(f20.2)', uf(0)
                return
                end
c
c=========================================
c
                subroutine outp(deltat)
c
c----------------------------------------------------------------------
                common/soln/ um(0:1880), uf(0:1880), c(0:1880,0:1880),
             $        sm(0:1880), sf(0:1880), p, msupp, n, t
                common/um0com/ num0, um0wid, um01(0:25), um02(0:25)
                common/uf0com/ nuf0, uf0wid, uf01(0:25), uf02(0:25)
                common/c0com/ nc0, c0wid, c01(0:25,0:25), c02(0:25,0:25)
                dimension value(5)
                data width /5.0/
                real solm(0:25), solf(0:25)
c
                nsum   = max (1, ifix(width / deltat))
                width  = nsum * deltat
                nwrite = msupp / nsum + 1
                nlines = (nwrite + width - 1) / width
                deltsq = deltat * deltat
```

```
c
550           format ('Step =',i4,5x,'Time =',f11.6,5x,'Max support =',i4,5x
     $           'Width =',f7.3)
              write (6,550) n, t, msupp, width
c
600           format ('POPULATION:',f14.2)
              write (6,600) p
c
              ii = 0
750           format ('DISTRIBUTION OF MALES:')
              write (6,750)
              nunwr = nwrite
              kmax  = -1
              do 10 l = 1,nlines
                mmax = min(5, nunwr)
                do 30 m = 1,mmax
                  value(m) = 0.0
                  kmin = kmax + 1
                  kmax = min(kmin + nsum - 1, msupp)
                  do 20 k = kmin,kmax
                    value(m) = value(m) + um(k)
20            continue
                  value(m) = value(m) * deltat
                  ii = (l-1)*5+m-1
                  solm(ii) = value(m)
30            continue
              write(6,1000) (value(m),m=1,mmax)
              nunwr = nunwr - 5
10            continue
                iim = ii
              print *,'ERROR IN THE COMPUTATION OF MALES:'
              if ((iim).lt.(num0)) then
                inum0 = num0 +1
                do 11 k = inum0,iim
                  solm(num0) = solm(num0) + solm(k)
11            continue
              else
                if ((iim).gt.(num0)) print*,'something is wrong'
              endif
              do 12 k = 0,num0-1
                umt02 = um02(k) * um0wid
                aerr = umt02 - solm(k)
                if(umt02.ne.0.0) rerr = aerr/umt02
                print '(3f20.2,f20.15)', umt02, solm(k), aerr, rerr
12            continue
```

```
c
 850          format ('DISTRIBUTION OF FEMALES:')
              write (6,850)
              nunwr = nwrite
              kmax  = -1
              do 40 l = 1,nlines
                mmax = min(5, nunwr)
                do 60 m = 1,mmax
                  value(m) = 0.0
                  kmin = kmax + 1
                  kmax = min(kmin + nsum - 1, msupp)
                  do 50 k = kmin,kmax
                    value(m) = value(m) + uf(k)
 50           continue
                  value(m) = value(m) * deltat
                  ii = (l-1)*5+m-1
                  solf(ii) = value(m)
 60           continue
              write(6,1000) (value(m),m=1,mmax)
              nunwr = nunwr - 5
 40           continue
                iif = ii
              print *,'ERROR IN THE COMPUTATION OF FEMALES:'
              if ((iif).lt.(nuf0)) then
                inuf0 = nuf0 +1
                do 13 k = inuf0,iif
                  solf(nuf0) = solf(nuf0) + solf(k)
 13           continue
              else
                if ((iif).gt.(nuf0)) print*,'something is wrong'
              endif
              do 14 k = 0,nuf0-1
                uft02 = uf02(k) * uf0wid
                aerr = uft02 - solf(k)
                if(uft02.ne.0.0) rerr = aerr/uft02
                print '(3f20.2,f20.15)', uft02, solf(k), aerr, rerr
 14           continue
c
 650          format ('COUPLES DISTRIBUTION:')
                        write (6,650)
              open(unit=1234,file='couples.d')
                        jmax = -1
                        sum = 0.0
                        do 100 lj = 1,nwrite
                          jmin = jmax + 1
                          jmax = min(jmin + nsum - 1, msupp)
```

```
                  nunwr = nwrite
                  imax = -1
                  do 70 li = 1,nlines
                    mmax = min(5, nunwr)
                    do 90 m = 1,mmax
                      value(m) = 0.0
                      imin = imax + 1
                      imax = min(imin + nsum - 1, msupp)
                      do 80 i = imin,imax
                      do 80 j = jmin,jmax
                        value(m) = value(m) + c(i,j)
   80           continue
                      value(m) = value(m) * deltsq
   110          format(i16)
                      write(1234,110) ifix(value(m))
   90           continue
                    write(6,1000) (value(m),m=1,mmax)
                    do 91 iii = 1,mmax
                      sum = sum + value(iii)
   91           continue
                    nunwr = nunwr - 5
   70           continue
                  write(6,*)
   100          continue
                close(1234)
c
   950          format ('-------------------------------',
           $       '-------------------------------')
                write(6,950)

                print *, ' SUM of all couples:'
                print '(f20.2)', sum
c
   1000         format (5(3x,f13.2))
c
                return
                end
c
c==========================================================
```

Bibliography

[1] T. Arbogast and F. A. Milner, *A finite difference method for a two-sex model of population dynamics*, SIAM J. Numer. Anal., 26 (1989), pp. 1474–1486.

[2] R. B. Banks, *Growth and Diffusion Phenomena*, Texts in Applied Mathematics, 14, Springer-Verlag, Berlin, 1994.

[3] N. Bennett, D. Bloom, and P. Craig, *American marriage patterns in transition*, in The Changing American Family: Sociological and Demographic Perspectives, S. J. South and S. Tolnay, eds., Westview, Boulder, CO, 1992, pp. 89–108.

[4] N. Bennett, D. Bloom, and C. Miller, *The influence of nonmarital childbearing on the formation of first marriages*, Demography, 32 (1995), pp. 47–62.

[5] H. Bernardelli, *Population waves*, J. Burma Res. Soc., 31 (1941), pp. 1–18.

[6] S. Blythe, C. Castillo-Chávez, and G. Casella, *Empirical methods for the estimation of the mixing probabilities for socially structured populations from a single survey sample*, Math. Population Studies, 3 (1992), pp. 199–225.

[7] S. Blythe, C. Castillo-Chávez, J. S. Palmer, and M. Cheng, *Toward a unified theory of sexual mixing and pair formation*, Math. Biosci., 107 (1991), pp. 379–405.

[8] R. Böckh, *Statistisches Jahrbuch der Stadt Berlin*, Statistik des Jahres 1884, Vol. 12, Berlin, 1886.

[9] S. Busenberg and C. Castillo-Chávez, *Interaction, pair formation and force of infection terms in sexually transmitted diseases*, in Mathematical and Statistical Approaches to AIDS Epidemiology, Lecture Notes in Biomathematics, 83, Springer-Verlag, Berlin, Heidelberg, New York, 1989, pp. 289–300.

[10] S. Busenberg and M. Iannelli, *Separable models in age-dependent population dynamics*, J. Math. Biol., 22 (1985), pp. 145–173.

[11] C. Castillo-Chávez and S. Blythe, *Mixing framework for social/sexual behavior*, in Mathematical and Statistical Approaches to AIDS Epidemiology, Lecture Notes in Biomathematics, 83, Springer-Verlag, Berlin, Heidelberg, New York, 1989, pp. 275–287.

[12] C. Castillo-Chávez and S. Busenberg, *On the solution of the two-sex mixing problem*, in Differential Equations Models in Biology, Epidemiology and Ecology, Lecture Notes in Biomathematics, 92, Springer-Verlag, Berlin, Heidelberg, New York, 1991, pp. 81–98.

[13] C. Castillo-Chávez, S. Busenberg, and K. Gerow, *Pair formation in structured populations*, in Differential Equations with Applications in Biology, Physics, and Engineering, Lecture Notes in Pure and Applied Mathematics, 133, Marcel Dekker, New York, 1991, pp. 47–65.

[14] C. Castillo-Chávez and W. Huang, *The logistic equation revisited*, Math. Biosci., 128 (1995), pp. 299–316.

[15] C. Castillo-Chávez, W. Huang, and J. Li, *On the existence of stable pairing distributions*, J. Math. Biol., 34 (1996), pp. 413–441.

[16] C. Castillo-Chávez, S.-F. Shyu, G. Rubin, and D. Umbach, *On the estimation problem of mixing/pair formation matrices with application to models for sexually-transmitted diseases*, in AIDS Epidemiology: Methodological Issues, N. Jewel, K. Dietz, and V. Farewell, eds., Birkhäuser, Boston, Basel, Berlin, 1992, pp. 384–402.

[17] C. Castillo-Chávez, A.-A. Yakubu, H. R. Thieme, and M. Martcheva, *Nonlinear mating models for populations with discrete generation*, in Mathematical Approaches for Emerging and Reemerging Infectious Diseases: An Introduction, C. Castillo-Chávez, S. Blower, P. Van den Driessche, D. Kirschner, and A.-A. Yakubu, eds., IMA, 125, Springer, New York, 2002, pp. 251–268.

[18] H. Caswell, *Matrix Population Models*, Sinauer Associates, Sunderland, MA, 1989.

[19] H. Caswell, *Matrix Population Models: Construction, Analysis and Interpretation* (second edition), Sinauer Associates, Sunderland, MA, 2000.

[20] A. J. Coale, *A New method for calculating Lotka's r—the intrinsic rate of growth in a stable population*, Population Studies, 11 (1957), pp. 92–94.

[21] A. J. Coale, *Age patterns of marriage*, Population Studies, 25 (1971), pp. 193–214.

[22] A. J. Coale, *The Growth and Structure of Human Populations, A Mathematical Investigation*, Princeton University Press, Princeton, NJ, 1972.

[23] A. J. Coale and D. R. McNeil, *The distribution by age of the frequency of first marriage in a female cohort*, J. Amer. Stat. Assoc., 67 (1972), pp. 743–749.

[24] P. Das Gupta, *On two-sex models leading to stable populations*, Theoret. Population Biol., 3 (1972), pp. 358–375.

[25] P. Das Gupta, *Growth of U.S. population*, 1940–1971, *in the light of an interactive two-sex model*, Demography, 10 (1973), pp. 543–565.

[26] P. Das Gupta, *An interactive nonrandom-mating two-sex model whose intrinsic growth rate lies between one-sex rates*, Theoret. Population Biol., 9 (1976), pp. 46–57.

[27] P. Das Gupta, *An alternative formulation of the birth function in a two-sex model*, Population Studies, 32 (1978), pp. 367–379.

[28] E. S. Deevey, Jr., *Life tables for natural populations of animals*, Quart. Rev. Biol., 22 (1947), pp. 283–314.

[29] K. Dietz and K. Hadeler, *Epidemiological models for sexually transmitted diseases*, J. Math. Biol., 26 (1988), pp. 1–25.

[30] L. Euler, *A general investigation into the mortality and multiplication of the human species*, Histoire de l'Académie Royale des Sciences et Belles Lettres, 16 (1760), pp. 144–164 (in French); translated by N. Keyfitz and B. Keyfitz, Theoret. Population Biol., 1 (1970), pp. 307–314.

[31] W. Feller, *On the integral equation of renewal theory*, Ann. Math. Stat., 12 (1941), pp. 243–267.

[32] A. Fredrickson, *A mathematical theory of age structure in sexual populations: Random mating and monogamous marriage models*, Math. Biosci., 20 (1971), pp. 117–143.

[33] B. Gompertz, *On the nature of the function expressive of the law of human mortality*, Philos. Trans., 27 (1825), pp. 513–585.

[34] L. A. Goodman, *Population growth of the sexes*, Biometrics, 9 (1953), pp. 212–225.

[35] L. A. Goodman, *On the reconciliation of mathematical theories of population growth*, J. Royal Stat. Soc., 130 (1967), pp. 541–553.

[36] L. A. Goodman, *On the age-sex composition of the population that would result from given fertility and mortality conditions*, Demography, 4 (1967), pp. 423–441.

[37] M. E. Gurtin and R. C. MacCamy, *Nonlinear age-dependent population dynamics*, Arch. Rat. Mech. Anal., 54 (1974), pp. 281–300.

[38] K. Hadeler, *Pair formation in age structured populations*, Acta Appl. Math., 14 (1989), pp. 91–102.

[39] K. Hadeler, *Structured population models for HIV infection pair formation and non-constant infectivity*, in AIDS Epidemiology: Methodological Issues, N. Jewel, K. Dietz, and V. Farewell, eds., Birkhäuser, Boston, Basel, Berlin, 1992, pp. 156–173.

[40] K. Hadeler, *Periodic solutions of homogeneous equations*, J. Differential Equations, 95 (1992), pp. 183–202.

[41] K. Hadeler, *Pair formation with maturation period*, J. Math. Biol., 32 (1993), pp. 1–15.

[42] K. HADELER AND K. NGOMA, *Homogeneous models for sexually transmitted diseases*, Rocky Mt. J. Math., 20 (1990), pp. 635–649.

[43] K. HADELER, R. WALDSTÄTTER, AND A. WÖRZ-BUSEKROS, *Models for pair formation in bisexual populations*, J. Math. Biol., 26 (1988), pp. 635–649.

[44] J. HAJNAL, *Some comments on Mr. Karmel's paper "The relations between male and female reproduction rates,"* Population Studies, 2 (1948), pp. 354–360.

[45] A. HASTINGS, *Population Biology: Concepts and Models*, Springer, New York, 1997.

[46] J. A. P. HEESTERBEEK AND J. A. J. METZ, *The saturating contact rate in marriage and epidemic models*, J. Math. Biol., 31 (1993), pp. 529–539.

[47] L. HENRY, *Nuptiality*, Theoret. Population Biol., 3 (1972), pp. 135–152.

[48] F. HOPPENSTEADT, *Mathematical Theories of Populations: Demographics, Genetics and Epidemics*, SIAM, Philadelphia, 1975.

[49] P. HSU AND A. FREDRICKSON, *Population-changing processes and the dynamics of sexual populations*, Math. Biosci., 26 (1975), pp. 55–78.

[50] S.-F. HSU SCHMITZ, *Some Theories, Estimation Methods and Applications of Marriage Functions Demography and Epidemiology*, Ph.D. dissertation, Cornell University, Ithaca, NY, 1994.

[51] S.-F. HSU SCHMITZ AND C. CASTILLO-CHÁVEZ, *Parameter estimation in non-closed social networks related to dynamics of sexually transmitted diseases*, in Modeling the AIDS Epidemic: Planning, Policy and Prediction, E. Kaplan and M. Brandeau, eds., Raven Press, New York, 1994, pp. 533–559.

[52] S.-F. HSU SCHMITZ AND C. CASTILLO-CHÁVEZ, *Completion of mixing matrices for non-closed social networks*, in Proceedings of the First World Congress of Nonlinear Analysis, Tampa, FL, August 19–26, 1992, Walter de Gruyter & Co., Berlin, 1996, pp. 3163–3173.

[53] S.-F. HSU SCHMITZ AND C. CASTILLO-CHÁVEZ, *On the evolution of age-structured marriage functions: It takes two to tango*, in Structured-Population Models in Marine, Terrestrial, and Freshwater Systems, S. Tuljapurkar and H. Caswell, eds., Chapman & Hall, New York, 1997, pp. 533–553.

[54] M. IANNELLI, *Mathematical Theory of Age-Structured Population Dynamics*, Comitato Nazionale per le Scienze Matematiche, C.N.R., 7, Giardini Editori e Stampatori, Pisa, 1995.

[55] M. IANNELLI AND M. MARTCHEVA, *A semigroup approach to the well posedness of the two-sex population model*, Dyn. Systems Appl., 6 (1997), pp. 353–370.

[56] M. IANNELLI AND F. MILNER, *On the approximation of the Lotka–McKendrick equation with finite life-span*, J. Comput. Appl. Math., 136 (2001), pp. 245–254.

[57] J. IMPAGLIAZZO, *Deterministic Aspects of Mathematical Demography*, Springer-Verlag, Berlin, Heidelberg, New York, Tokyo, 1985.

[58] H. INABA, *Weak ergodicity of population evolution process*, Math. Biosci., 96 (1985), pp. 195–219.

[59] H. INABA, *Human population reproduction via first marriage*, Math. Population Studies, 5 (1995), pp. 123–144.

[60] H. INABA, *Persistent age distributions for an age-structured two-sex population model*, Math. Population Studies, 7 (2000), pp. 365–398.

[61] P. H. KARMEL, *The relations between male and female reproduction rates*, Population Studies, 1 (1947), pp. 249–274.

[62] P. H. KARMEL, *The relations between male and female nuptiality in a stable population*, Population Studies, 1 (1948), pp. 353–387.

[63] P. H. KARMEL, *An analysis of the sources and magnitudes of inconsistencies between male and female net reproductive rates in actual populations*, Population Studies, 2 (1948), pp. 240–273.

[64] P. H. KARMEL, *A rejoinder to Mr. Hajnal's comments*, Population Studies, 2 (1948), pp. 361–372.

[65] D. G. KENDALL, *Stochastic processes and population growth*, J. Royal Stat. Soc. Ser. B, 11 (1949), pp. 230–264.

[66] N. KEYFITZ, *Reconciliation of population models: Matrix integral equation and partial fraction*, J. Royal Stat. Soc., 130 (1967), pp. 61–83.

[67] N. KEYFITZ, *Introduction to Mathematics of Population*, Addison-Wesley, Reading, MA, 1968.

[68] N. KEYFITZ, *Changing vital rates and age distributions*, Population Studies, 22 (1968), pp. 235–251.

[69] N. KEYFITZ, *The mathematics of sex and marriage*, in Proceedings of the Sixth Berkeley Symposium on Mathematical Statistics and Probability VI: Biology and Health, University of California Press, Berkeley, CA, 1972, pp. 89–108.

[70] N. KEYFITZ, *Applied Mathematical Demography*, Springer-Verlag, New York, Berlin, Heidelberg, Tokyo, 1985.

[71] M. KOT, *Elements of Mathematical Ecology*, Cambridge University Press, Cambridge, UK, 2001.

[72] H. LANGHAAR, *General population theory in the age-time continuum*, J. Franklin Inst., 293 (1972), pp. 199–214.

[73] P. H. LESLIE, *On the use of matrices in certain population mathematics*, Biometrika, 35 (1945), pp. 183–212.

[74] E. G. LEWIS, *On the generation and growth of a population*, Sankhyā, 6 (1942), pp. 93–96.

[75] L. LILLARD AND C. PANIS, *Marital status and mortality: The role of health*, Demography, 33 (1996), pp. 313–327.

[76] A. LOTKA, *Relation between birth rates and death rates*, Sci. N. S., 26 (1907), pp. 21–22.

[77] A. LOTKA, *The stability of the normal age distribution*, Proc. Nat. Acad. Sci. USA, 8 (1922), pp. 339–345.

[78] A. J. LOTKA, *Analytical Theory of Biological Populations*, Plenum Press, New York, 1998.

[79] W. M. MAKEHAM, *On the law of mortality and the construction of annuity tables*, Assurance Magazine, 8 (1860), pp. 301–310.

[80] W. M. MAKEHAM, *On the law of mortality*, J. Inst. Actuaries, 13 (1867), pp. 325–367.

[81] TH. R. MALTHUS, *An Essay on the Principle of Population*, printed for J. Johnson, St. Paul's church-yard, London, 1798; also available online from http://www.ac.wwu.edu/~stephan/malthus/malthus.0.html.

[82] M. MARTCHEVA, *Exponential growth in age-structured two-sex populations*, Math. Biosci., 157 (1999), pp. 1–22.

[83] M. MARTCHEVA AND F. MILNER, *A two-sex age-structured population model: Well posedness*, Math. Population Studies, 7 (1999), pp. 111–129.

[84] M. MARTCHEVA AND F. MILNER, *The mathematics of sex and marriage, revisited*, Math. Population Studies, 9 (2001), pp. 123–141.

[85] T. MATSUMOTO, S. OHARU, AND H. R. THIEME, *Nonlinear perturbations of a class of integrated semigroups*, Hiroshima Math. J., 26 (1996), pp. 433–473.

[86] J. C. MCCANN, *A more accurate short method of approximating Lotka's r*, Demography, 10 (1973), pp. 567–570.

[87] D. MCFARLAND, *Comparison of alternative marriage models*, in Population Dynamics, T. N. T. Greville, ed., Academic Press, New York, London, 1972, pp. 89–106.

[88] D. MCFARLAND, *Models of marriage formation and fertility*, Social Forces, 54 (1975), pp. 66–83.

[89] A. C. MCKENDRICK, *Applications of mathematics to medical problems*, Proc. Edinburgh Math. Soc., 44 (1926), pp. 98–130.

[90] F. A. MILNER, *A finite element method for a two-sex model of population dynamics*, Numer. Methods Partial Differential Equations, 4 (1988), pp. 329–345.

[91] F. A. Milner and G. Rabbiolo, *Rapidly converging numerical algorithms for models of population dynamics*, J. Math. Biol., 30 (1992), pp. 733–753.

[92] S. Mitra, *Effect of adjustment for the sex composition in the measurement of fertility on the intrinsic rates*, Demography, 13 (1976), pp. 251–257.

[93] S. Mitra, *On the derivation of a two-sex stable model*, Demography, 15 (1978), pp. 541–548.

[94] S. Mitra, *On Das Gupta's "Birth function in a two-sex model,"* Population Studies, 34 (1980), pp. 566–567.

[95] National Center for Health Statistics, *National Vital Statistics Reports*, 1989–90, Advance Report of Final Divorce Statistics, 1989 and 1990, Vol. 43, No. 9, Public Health Service, Centers for Disease Control and Prevention, Atlanta, GA, 1995.

[96] National Center for Health Statistics, *National Vital Statistics Reports*, 2000, Births: Final Data for 2000, Vol. 50, No. 5, Public Health Service, Centers for Disease Control and Prevention, Atlanta, GA, 2002.

[97] National Center for Health Statistics, *National Vital Statistics Reports*, 2000, Death: Final Data for 2000, Vol. 50, No. 15, Public Health Service, Centers for Disease Control and Prevention, Atlanta, GA, 2002.

[98] B. Parlett, *Can there be a marriage function?*, in Population Dynamics, T. N. T. Greville, ed., Academic Press, New York, London, 1972, pp. 107–135.

[99] R. Pollak, *A reformulation of the two-sex problem*, Demography, 23 (1986), pp. 247–259.

[100] R. Pollak, *The two-sex model with persistent unions: A generalization of the birth matrix-mating rule model*, Theoret. Population Biol., 32 (1987), pp. 176–187.

[101] R. Pollak, *Two-sex demographic model*, J. Polit. Econ., 98 (1990), pp. 399–420.

[102] R. Pollak, *Two-sex population models and classical stable population theory*, in Convergent Issues in Genetics and Demography, J. Adams, D. Lam, A. Hermalin, and P. Smouse, eds., Oxford University Press, Oxford, UK, 1990, pp. 317–333.

[103] A. H. Pollard, *The measurement of reproductivity*, J. Inst. Actuaries, 74 (1948), pp. 288–318.

[104] J. H. Pollard, *A Taylor series expansion for Lotka's r*, Demography, 7 (1970), pp. 151–154.

[105] J. H. Pollard, *Mathematical Models for the Growth of Human Populations*, Cambridge University Press, Cambridge, UK, 1973.

[106] J. H. Pollard, *Modelling the interaction between the sexes*, Math. Comput. Modelling, 26 (1997), pp. 11–24.

[107] J. Prüss and W. Schappacher, *Persistent age-distributions for a pair-formation model*, J. Math. Biol., 33 (1994), pp. 17–33.

[108] A. Pugliese, *Contact matrices for multipopulation epidemic models: How to build a consistent matrix close to data*, IMA J. Math. Appl. Med. Biol., 8 (1991), pp. 249–271.

[109] R. Schoen, *A two-sex nuptiality-mortality life table*, Demography, 14 (1977), pp. 333–350.

[110] R. Schoen, *The harmonic mean as the basis of a realistic two-sex marriage model*, Demography, 18 (1981), pp. 201–216.

[111] R. Schoen, *Generalizing the life table model to incorporate interactions between the sexes*, in Multidimensional Mathematical Demography, K. C. Land and A. Rogers, eds., Academic Press, New York, 1982.

[112] R. Schoen, *Modeling Multigroup Populations*, Plenum Press, New York, London, 1988.

[113] R. Schoen and R. Weinick, *The slowing metabolism of marriage: Figures from* 1988 *U.S. marital status life tables*, Demography, 30 (1993), pp. 734–746.

[114] F. Sharpe and A. Lotka, *A problem in age distribution*, Philos. Mag., Ser. 6, 21 (1911), pp. 435–438.

[115] E. Sinestrari, *Non-linear age-dependent population growth*, J. Math. Biol., 9 (1980), pp. 331–345.

[116] D. Smith and N. Keyfitz, *Mathematical Demography: Selected Papers*, Biomathematics, 6, Springer-Verlag, New York, 1977.

[117] S. South, *For love or money? Sociodemographic determinants of the expected benefits from marriage*, in The Changing American Family: Sociological and Demographic Perspectives, S. J. South and S. Tolnay, eds., Westview, Boulder, CO, 1992, pp. 171–194.

[118] U.S. Bureau of the Census, *Census of Population:* 1970, Marital Status, Final Report PC(2)-4C, U.S. Summary, U.S. Government Printing Office, Washington, DC, 1972.

[119] U.S. Bureau of the Census, *Census of Population:* 1970, Detailed Characteristics, Final Report PC(1)-D1, U.S. Summary, U.S. Government Printing Office, Washington, DC, 1973.

[120] U.S. Bureau of the Census, *Census of Population:* 1980, Vol. 1, Characteristics of Population, PC 80-1-B1, U.S. Government Printing Office, Washington, DC, 1981.

[121] U.S. Bureau of the Census, *Census of Population:* 1980, Vol. 2, Subject Reports: Marital Characteristics, PC 80-2-4C, U.S. Government Printing Office, Washington, DC, 1984.

[122] U.S. Department of Health and Human Services, *Vital Statistics of the United States,* 1990, Vol. II, DHHS Publication No. (PHS) 94-1104, Public Health Service, Centers for Disease Control and Prevention, Hyattsville, MD, 1994.

[123] P.-F. Verhulst, *A note on the law of population growth*, Correspondence Mathématique et Physique Publée par A. Quételet, 10 (1838), pp. 113–121.

[124] H. von Foerster, *Some remarks on changing populations*, in The Kinetics of Cellular Proliferation, F. Stohlman, Jr., ed., Grune and Stratton, New York, 1959, pp. 382–407.

[125] G. Webb, *Theory of Nonlinear Age Dependent Population Dynamics*, Marcel Dekker, New York, 1985.

[126] G. Webb, *A semigroup proof of the Sharpe–Lotka theorem*, in Infinite-Dimensional Systems, Lecture Notes in Mathematics, 1076, Springer-Verlag, Berlin, New York, 1994, pp. 254–268.

[127] J. Yellin and P. Samuelson, *A dynamical model for human population*, Proc. Nat. Acad. Sci. USA, 71 (1974), pp. 2813–2817.

[128] J. Yellin and P. Samuelson, *Comparison of linear and nonlinear models for human population dynamics*, Theoret. Population Biol., 11 (1977), pp. 105–126.

[129] R. Zacher, *Persistent solutions for age-dependent pair-formation models*, J. Math. Biol., 42 (2001), pp. 507–531.

Index